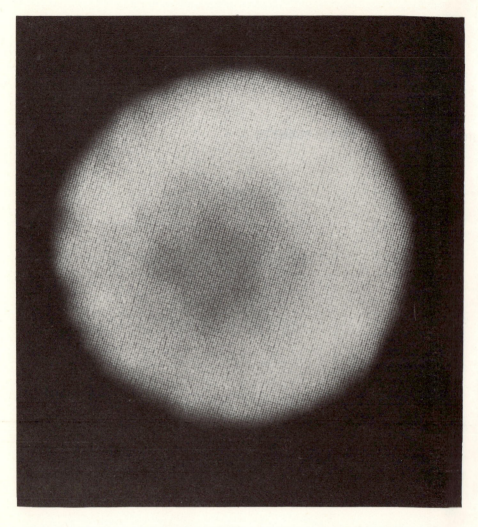

This image of Uranus was obtained by B.A. Smith using a charge-coupled device (CCD) at the 1.54m Catalina telescope of the University of Arizona. A filter was used to isolate light in the 890 nm infrared absorption band of methane, a wavelength region in which methane gas strongly absorbes light. The bright limb of Uranus at this wavelength is evidence of a hazy layer of ice crystals high in the atmosphere of the giant planet. This image was made possible by the high infrared sensitivity of the CCD, which acts as a solid-state television-like imaging device.

Uranus
and the outer planets

Proceedings of the IAU/RAS colloquium no. 60

Edited by
GARRY HUNT
Director, Laboratory for Planetary Atmospheres
University College, London

CAMBRIDGE UNIVERSITY PRESS
Cambridge
London   New York   New Rochelle
Melbourne   Sydney

Published by the Press Syndicate of the University of Cambridge
The Pitt Building, Trumpington Street, Cambridge CB2 1RP
32 East 57th Street, New York, NY 10022, USA
296 Beaconsfield Parade, Middle Park, Melbourne 3206, Australia

First published 1982

Printed in Great Britain at the University Press, Cambridge

Library of Congress catalogue card number: 81-17047

British Library cataloguing in publication data

Uranus and the outer planets.
Uranus (Planet)
I. Hunt, Garry
523.4'7   QB681

ISBN 0 521 24573 7

CONTENTS

PREFACE

The IAU/RAS Colloquium No.60 entitled "Uranus and the Outer Planets" was held at the University of Bath during 14-16 April 1981. The meeting was attended by 164 scientists from 13 countries - USA, USSR, Germany, France, Italy, Brazil, Holland, Canada, Israel, Norway, Eire, Belgium and the UK - all of whom actively participated in the meeting, which covered the past, present and future observations of Uranus.

The meeting was held to celebrate the 200th anniversary of the discovery of the planet Uranus by Sir William Herschel, FRS, so that it was essential that the meeting be held in the city of Bath where the discovery was made. The organisation of the conference was carried out jointly with the Royal Astronomical Society, whose first President was Sir William Herschel FRS.

The Scientific Organising Committee consisted of G.E. Hunt (UK, Chairman), A. Brahic (France), R.D. Davies (UK), B.A. Smith (USA), D. Gautier (France), T. Kiang (Eire), T. Owen (USA), S.K. Runcorn (UK), B.A. Smith (USA), V.G. Tejfel (USSR) and A. Wosczyk (Poland). The local organisation for the meeting was carried out by R.D. Davies, A. Steff-Langston, and members of the staff at the Royal Astronomical Society, together with D. Parkin and his colleagues at the Department of Physics, University of Bath.

The scientific discussions were divided into six main sections with 17 invited and 22 contributed papers. The meeting started with a detailed history of astronomical studies by Herschel in the 18th century which led to this important discovery, and its effect upon astronomy at that time. Four sections covered our present knowledge of the planet, its atmosphere, rings and satellites. However, there are still exciting future prospects for the exploration of Uranus and its surrounding environment in the next few years, which is essential if we are to answer the major problems discussed at the meeting. The concluding discussions indicated the important studies of Uranus that will be possible with Space Telescope; also from the observations that will be made by Voyager during its encounter with Uranus in 1986; and from future space missions.

The Colloquium brought together a very wide collection of the scientific community to celebrate in Bath one of the most important discoveries in astronomy. Since Sir William Herschel FRS was a distinguished musician, the participants and their guests were also able to hear some of his music at a concert given by the Herschel Ensemble, performed in the attractive surroundings of the Guildhall. In addition, the complete dimensions of Sir William Herschel's talents were described in a public lecture given by Professor Sir Bernard Lovell FRS, who also added a further musical flavour to the astronomical proceedings.

This book contains all the invited lectures presented at the meeting. The written record of the meeting is introduced by Professor A. Wolfendale FRS, the current President of the Royal Astronomical Society. Herschel's original paper about Uranus has been reproduced from the Philosophical Transactions of The Royal Society, and we are very grateful to The Royal Society for their permission to include this important paper in this book. The historical sections provide the amusing and scientifically stimulating discussions relating to Herschel's work and the impact on astronomy of this important planetary discovery. In the section on the current knowledge about Uranus, the paper by Elliot also includes details of occultation observations made during April 1981. These discussions are therefore as up-to-date as possible.

The final discussions look towards the exciting future with the Voyager encounter and observations using Space Telescope, with a final contribution provided by Professor P. Wayman, Secretary General of the IAU.

This volume therefore provides a very important document about Uranus; not only reviewing our current knowledge, but also providing the detailed historical background to astronomy in the

18th Century.

I would like to express my thanks to all members of the
organising committee for their assistance in organising the meeting,
to the authors for promptly producing their manuscripts for this
volume, and to my secretary, Pamela Shapiro, for her invaluable
assistance.

Garry E. Hunt

Laboratory for Planetary Atmospheres
Department of Physics and Astronomy,
University College London

July 1981

INTRODUCTION

Crest of The Royal Astronomical Society. The instrument
shown is Sir William Herschel's 40 foot telescope. Herschel
was the first President of the R.A.S.

The Latin tag reads "Whatever shines is to be noticed".

URANUS AND THE OUTER PLANETS

INTRODUCTORY REMARKS

by

A.W. WOLFENDALE F.R.S.

UNIVERSITY OF DURHAM, U.K.

(PRESIDENT OF THE ROYAL ASTRONOMICAL SOCIETY)

It is a pleasure to write this short Introduction to the joint RAS/IAU Meeting on "Uranus and the Outer Planets". Sir William Herschel, who discovered the planet Uranus two hundred years ago in the city of Bath, was the first President of the Royal Astronomical Society. We are continually reminded of his work by our crest (Frontispiece) which shows Herschel's giant telescope – some forty feet in length and containing a four foot diameter mirror. It is a remarkable fact that this giant instrument came into use only eight years after he had taken up full time work in Astronomy and it remained the world's biggest for many years.

The important role of the amateur in the advancement of science has been well demonstrated, and Astronomy has had more than its share of gifted amateurs, but Herschel's contribution was quite outstanding. By concentrating effort on producing instruments of quality far above those available to contemporary amateurs (and, as it transpired, superior even to those used by professionals), and by having a lack of preconceived ideas as to what the Heavens should contain, Herschel was able to make what in modern jargon we would refer to as a quantum leap in the subject.

Herschel was 42 when he discovered what was later to be known as Uranus. Plate 1 shows him at a somewhat later age (56).

Some small idea of Herschel's constructional skill can be gained from Plate 2 which shows the telescope, currently in the Science Museum, London, which is a near-contemporary copy of the instrument used in the discovery of Uranus. Of course, the quality of the telescope was conditioned largely by the speculum mirrors, hand-ground and polished; it was these mirrors which even the Astronomer Royal of the time, Nevil Maskelyne, acknowledged to be superior to his own.

Later papers will deal with Herschel's contributions to planetary science in detail; here I will consider some other aspects of his work.

Although the majority of the credit for Herschel's discoveries must go to the man himself, some is due to his sister Caroline (Plate 3). Miss Herschel joined him in 1772, initially as a house-keeper, and, after tuition, became a painstaking and highly proficient observer in her own right. The RAS acknowledged Caroline's researches by the award of its Gold Medal in 1828 and I should like to refer to the address by the Vice President of the R.A.S. at the time (J. South) on the occasion of the announcement of the award. South mentions the key role played by Caroline Herschel in noting the measurements made by her brother at the telescope, of her reduction of the observations and of her planning of the surveys. The Vice President refers to

"the discovery of the comet of 1786, of the comet of
    1788, of the comet of 1791, of the comet of 1793,
    and of the comet of 1795, since rendered familiar
    to us by the remarkable discovery of Encke. Many also
    of the nebulae contained in Sir W. Herschel's
    catalogues were detected by her during these hours of
    enjoyment. Indeed, in looking at the joint labours of
    these extraordinary personages, we scarcely know
    whether most to admire the intellectual power of the
    brother, or the unconquerable industry of the sister."

Finally, South gives the following description of what was Caroline's most important contribution.

"Unwilling, however, to relinquish her astronomical
    labours whilst anything useful presented itself,
    she undertook and completed the laborious reduction
    of the places of 2,500 nebulae, to the 1st of
    January, 1800, presenting in one view the results
    of all Sir William Herschel's observations on those
    bodies, thus bringing to a close half a century spent
    in astronomical labour."

It was on these grounds then that Caroline Herschel was awarded the
Gold Medal.

Returning to William Herschel, himself, it is important to
stress that his contributions to Astronomy were by no means limited
to nearby objects. There were analyses of double stars and the
motion of the solar system with respect to the local stars, but
Herschel's achievement of greatest significance was in his development of theories to explain 'the construction of the heavens',
theories largely based on his own comprehensive observational data.
Observations of nebulae were crucial in this work and Herschel's
systematic 20-year observing programme gave rise to the list of
2,500 such nebulae, some 25 times the number previously known.
The incomparable quality of his telescopes led to the resolution
of many nebulae into individual stars and the realisation that
some comprised incandescent gas. Inevitably, there is a tangled
web of fact and interpretation but there can be traced in this
work the beginnings of contemporary views about the life history
of star clusters, the shape of our own Galaxy, the multiplicity
of galaxies and, not least, possible modes of generation of
galaxies.

ACKNOWLEDGEMENTS

The author wishes to thank Mrs. Enid Lake (Royal Astronomical
Society Library), Mr. N.H. Robinson (Royal Society Library),
Dr. J. Darius (Science Museum, London) and Mr. A.D. Burnett
(University Library, Durham) for their help.

PLATE 1   6

Sir William Herschel, F.R.S. (at the age of 56). From a
pastel by J. Russell R.A. 1794 (reproduced in "The Royal
Society/Royal Astronomical Society Collected Scientific
Papers of Sir Wm. Herschel").

PLATE 2   7

A contemporary copy of the 7 foot telescope with which
Herschel discovered Uranus on 1781 March 13 at Bath
(Science Museum, London).

PLATE 3

Caroline Herschel (at the age of 79). From the portrait
in oils by Tielmann in 1829 (reproduced in "The Royal
Society/Royal Astronomical Society Collected Scientific
Papers of Sir Wm. Herschel").

XXXII. *Account of a Comet.* *By Mr*. Herfchel, *F. R. S.;* *communicated by Dr*. Watfon, *Jun. of* Bath, *F. R. S.*

Read April 26, 1781.

ON Tuefday the 13th of March, between ten and eleven in the evening, while I was examining the fmall ftars in the neighbourhood of H Geminorum, I perceived one that appeared vifibly larger than the reft : being ftruck with its uncommon magnitude, I compared it to H Geminorum and the fmall ftar in the quartile between Auriga and Gemini, and finding it fo much larger than either of them, fufpe&ed it to be a comet.

I was then engaged in a feries of obfervations on the parallax of the fixed ftars, which I hope foon to have the honour of laying before the Royal Society; and thofe obfervations requiring very high powers, I had ready at hand the feveral magnifiers of 227, 460, 932, 1536, 2010, &c. all which I have fuccefsfully ufed upon that occafion. The power I had on when I firft faw the comet was 227. From experience I knew that the diameters of the fixed ftars are not proportionally magnified with higher powers, as the planets are; therefore I now put on the powers of 460 and 932, and found the diameter of the comet increafed in proportion to the power, as it ought to be, on a fuppofition of its not being a fixed ftar, while the diameters of the ftars to which I compared it were not increafed in the fame ratio. Moreover, the comet being magnified much beyond what its light would admit of, appeared hazy and ill-defined with thefe great powers, while the ftars preferved that luftre and diftinctnefs which from many thoufand obfervations I knew they would retain. The fequel has fhewn that my furmifes were well founded, this proving to be the Comet we have lately obferved.

I have reduced all my obfervations upon this Comet to the following tables. The firft contains the meafures of the gradual increafe of the Comet's diameter. The micrometers I ufed, when every circumftance is favourable, will meafure extremely fmall angles, fuch as do not exceed a few feconds, true to 6, 8, or 10 thirds at moft; and in the worft fituations true to 20 or 30 thirds: I have therefore given the meafures of the Comet's diameter in feconds and thirds. And the parts of my micrometer being thus reduced, I have alfo given all the reft of the meafures in the fame manner; though in large diftances, fuch as one, two, or three minutes, fo great an exactnefs, for feveral reafons, is not pretended to.

TABLE

TABLE I. Meafures of the Comet's diameter *.

Days.	″ ‴	Powers.
March 17	2 53	932. 460.
19	2 59	932. 460.
21	3 38	460.
28	4 7	932 ⎫ thefe meafures agree to 9‴.
—	3 58	227 ⎭
29	4 7	227 rather too fmall a meafure.
—	4 25	227 feems right.
April 2	4 25	227
6	4 53	227
15	5 11	227
—	5 20	227 very good; not liable to half a fecond of error.
18	5 2	227 true to 12″ or 18‴ at moft.

Having meafured the diameter of the Comet with fuch high power as 932 and 460, it may not be amifs to make one obfervation on this fubject, left it fhould be mifapprehended that I pretend to a diftinct power of fuch magnitude upon all celeftial objects in general. By experience I have found, that the aberration or indiftinctnefs occafioned by magnifying much, provided the object be ftill left fufficiently diftinct, is rather to be put up with, than the power to be reduced, when the angles to be meafured are extremely fmall. The reafon of this may, perhaps, be that a fmall error of judgement, to which we are always liable, is of great confequence with a low power, as bearing a confiderable proportion to the diameter of the object;

* There are feveral optical deceptions which may affect the meafures of objects that fubtend extremely fmall angles. Thus I have found, by experience, that a very fmall object will appear fomething lefs in a telefcope when we fee it firft than when we become familiar with it. There is alfo a deflection of light upon the wires when they are nearly fhut; but as none of thefe deceptions are well enough underftood to apply a correction, I leave them affected with them.

whereas

whereas with a higher power the proportion of this error to the whole becomes much lefs, and the meafure more exact, even after we have made allowance for a fmall additional error occafioned by the want of that perfect diftinctnefs which is required for other purpofes. However, to enter deeply into an explanation of this would lead me to fpeak of the caufes of the aberration of rays in the focus of an object fpeculum, of which there are fome that are feldom taken into confideration by opticians, and indeed are fuch as cannot be calculated; but this not being my prefent purpofe, fuffice it to obferve, that the method is juftified by experience.

When the diameter of the Comet was increafed to about 4″, I thought it advifable to leffen the power with which I meafured; and, as I made ufe of two different micrometers, as well as eye-glaffes, I took a meafure with both of them. The agreement of the micrometers to 9‴ is no fmall proof of the goodnefs of the obfervations of the 28th of March, and very properly connects the meafures of the high powers with thofe that were made with 227.

TABLE II. Diſtance of the Comet from certain teleſcopic fixed ſtars which I have marked α, β, γ, δ, ε, ζ.

D. H. M.		′ ″ ‴	
Mar.13 10 30	from α,	2 48 0	by pretty exact eſtimation true to 20″.
17 11 0	fig. 1*.	0 41 58	by the micrometer and power 227,
18 7 20		1 0 35	
— 9 16		1 6 59	
— 10 55		1 10 40	
19 7 4		1 46 40	
— 10 42		1 51 23	
21 10 0		3 39 46	
24 8 12	from β,	2 55 39	true to 4 or 5″, an indifferent obſervation.
— 10 58	fig. 2.	2 53 4	true to 4 or 5″.
25 7 24		2 12 46	true to 2 or 3″.
— 9 47		2 14 18	
26 10 43		1 48 3	true to 2 or 3″.
28 7 46		2 55 49	true to 4 or 5″.
29 8 50	from γ.	2 20 51	true to 2″.
30 7 55	fig. 3.	1 28 48	true to 2 or 3″.
Apr. 1 7 45		2 39 20	
6 8 50	from δ. fig. 4.	2 51 23	
15 10 18	from ε.	4 27 57	eſtimated by the field, true to 5 or 6″.
16 7 50	fig. 5.	3 9 14	by the micrometer, true to 3 or 4″.
— 10 47		2 50 56	true to 3 or 4″.
18 8 18		3 18 4 }	mean 3′ 17″, true to 1″ or 1½.
— —		3 15 57 }	
— 8 50	from ζ,	2 24 57	
19 8 38	fig. 6.	3 2 5	true to 3 or 4″.

* The figures are drawn upon a ſcale of 80 ſeconds to one inch.

TABLE

TABLE III. Angle of poſition of the Comet with regard to the parallel of declination of the ſame teleſcopic fixed ſtars meaſured by a micrometer, of which I have given the deſcription, and a magnifying power of 278. See fig. 1. 2. 3. 4. 5. 6.

D. H. M.		° ′	
Mar.13 10 30	Bα Comet,	0 0	} by ſuperficial eſtimation, liable to an error of 10 or 12 degrees.
17 11 0	Aα Comet,	89 56	by the micrometer.
18 8 20	fig. 1 *.	56 39	
— 9 24		41 33	true to 1°.
19 7 23		29 47	true to 1°.
21 10 10		11 46	true to 4 or 5°.
— 11 48		12 14	
24 8 23	Bβ Comet,	38 39	true to 2 or 3°.
— 11 4	fig. 2.	36 14	true to 3 or 4°, air very tremulous.
25 7 33		53 18	
— 9 55		56 32	liable to a conſiderable error.
26 10 55	Aβ Comet.	87 0	true to 2 or 3°.
28 7 58		28 51	true to 3 or 4°.
29 9 25	Bγ Comet,	32 19	true to 1 or 2°.
30 8 25	fig. 3.	72 14	true to 3 or 4°.
Apr. 1 7 55	Aγ Comet,	28 51	well taken, } 27° 46′, true to 1°.
— — —		27 14	more exact,
6 8 28	Bδ Comet, fig. 4.	84 42	true to leſs than 2°
15 10 27	Bε Comet,	29 9	true to 2 or 3°.
16 8 1	fig. 5.	49 11	true to 1°.
— 10 55		50 47	true to 1½ or 2°½.
18 8 31	Aε Comet,	47 9	very well taken, } 47°, true to leſs than 1°.
— — —		46 35	pretty well,
— 9 8	Bζ Comet,	82 39	
19 8 56	Aζ Comet,	48 18 } 49° 3′, true to 1°.	
— — —	fig. 6.	49 48	
— 10 45		47 30	true to 2 or 3°.

* The angles are drawn true to the meaſure, without allowing for errors.

Miſcellaneous obſervations and remarks.

March 19. The Comet's apparent motion is at preſent 2¼ ſeconds *per* hour. It moves according to the order of the ſigns, and its orbit declines but very little from the ecliptic.

March 25. The apparent motion of the Comet is accelerating, and its apparent diameter ſeems to be increaſing.

March 28. The diameter is certainly increaſed, from which we may conclude that the Comet approaches to us.

April 2. This evening at 8 h. 15′ the Comet was a little above the line drawn from η to θ in fig. 7. This figure is only delineated by the eye, ſo that no very great exactneſs in the diſtances of the ſtars is to be expected; but I ſhall take the firſt opportunity of meaſuring their reſpective ſituations by the micrometer.

April 6. With a magnifying power of 278 times the Comet appeared perfectly ſharp upon the edges, and extremely well defined, without the leaſt appearance of any beard or tail.

April 16. Fig. 8. repreſents the ſituation of the Comet this evening about nine o'clock, and is only an eye-draught of the teleſcopic ſtars.

Remarks on the path of the Comet.

We may obſerve, that the method of tracing out the path of a celeſtial body by taking its diſtance from certain ſtars, and the angle of poſition with regard to them, cannot be expected to give us a compleatly juſt repreſentation of the tract it deſcribes, ſince even the moſt careful obſervations are liable to little errors, both from the remaining imperfections of inſtruments, though

they

they fhould be the moft accurate that can be had, and from the difficulty of taking angles and pofitions of objects in motion. Add to this a third caufe of error, namely, the obfcurity of very fmall telefcopic ftars that will not permit the field of view fo well to be enlightened as we could wifh, in order to fee the threads of the micrometer perfectly diftinct.

This will account for the apparent diftortions to be obferved in my figures of the Comet's path. Some little irregularity therein may alfo proceed from different refractions, as they have not been taken into account, though the obfervations have been made at very different altitudes, where confequently the refractions muft have been very different. But though this method may be liable to great inconveniences, the principal of which is, that many parts of the heavens are not fufficiently ftored with fmall ftars to give us an opportunity to meafure from them, yet the advantages are not lefs remarkable. Thus we fee that it enabled me to diftinguifh the quantity and direction of the motion of this Comet in a fingle day (from the 18th to the 19th of March) to a much greater degree of exactnefs than could have been done in fo fhort a time by a fector or tranfit inftrument; nay even an hour or two, we fee, were intervals long enough to fhew that it was a moving body, and confequently, had its fize not pointed it out as a Comet, the change of place, though fo trifling as $2\frac{1}{4}$ feconds *per* hour, would have been fufficient to occafion the difcovery. A gentleman very well known for his remarkable fuccefs in detecting Comets * feems to be well aware of the difficulty to difcover a motion in a heavenly body by the common methods when it is fo very fmall; for in a letter he favoured me with, fpeaking of the Comet, he fays: " Rien n'etoit plus difficile que de la " reconnoître et je ne puis pas concevoir comment vous avés pu

* Monf. MESSIER.

" revenir

" revenir plufieurs fois fur cette étoile ou Comête ; car abfolu-
" ment il a fallu l'obferver plufieurs jours de fuîte pour s'ap-
" perçevoir qu'elle avoit un mouvement."

I need not fay that I merely point this out as a temporary advantage in the method I have taken ; for as foon as we can have regular, conftant, and long continued obfervations by fixed inftruments, the excellence of them is too well known to fay any thing upon that fubject : for which reafon I failed not to give immediate notice of this moving ftar, and was happy to furrender it to the care of the Aftronomer Royal and others, as foon as I found they had begun their obfervations upon it.

Facsimile of Herschel's paper kindly provided by the Library, Institute of Astronomy, University of Cambridge

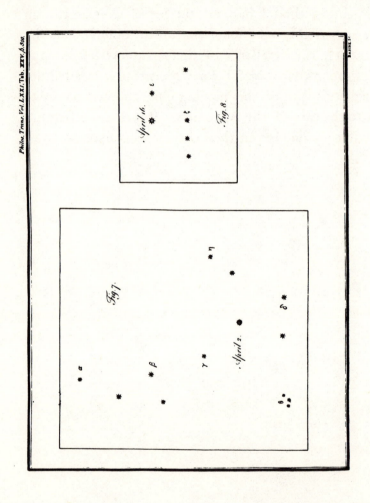

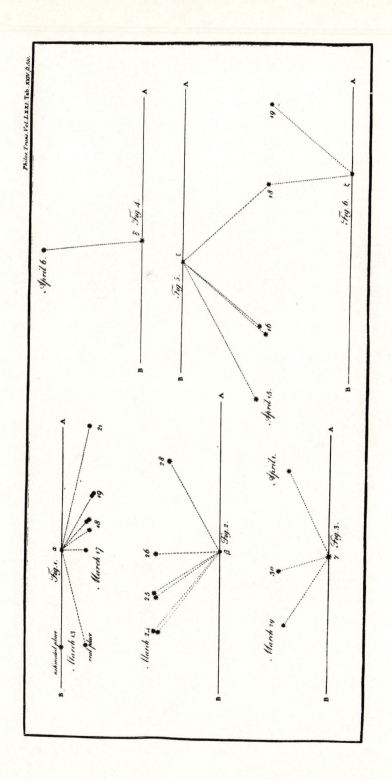

HISTORY OF THE DISCOVERY OF URANUS

WILLIAM HERSCHEL, BATH, and the PHILOSOPHICAL SOCIETY

by

Roy Porter

Wellcome Institute for the
History of Medicine, London

Richard Brinsley Sheridan, a Bathonian, opens his
comedy, *The Rivals*, with a social sneer. Fag, the stylish
apostle of *bon ton*, meets his fellow servant Thomas, a clod-
hopping yokel, and exclaims: "Why Thomas ... But who the
deuce thought of seeing *you* in Bath?". One might ask the
same of William Herschel, the Hanoverian bandsman's son, who,
having discovered Uranus, was to pass the last forty years of
his life as a Crown pensionary, living just outside Windsor.
How did Herschel come to spend his formative scientific years
in Bath, resort of the giddy and the gay? How did he come
to discover Uranus there?

It was not as an astronomer that Herschel moved to
Bath. He earned his living as a musician. Son of a German
regimental bandmaster, William came to England in 1757 with
his oboe to try to make his fortune through his musical
talents. Finding London overstocked with musicians, he
migrated to the North East, where, in towns such as Richmond,
Newcastle, Leeds and Pontefract, he pieced together a liveli-
hood by playing in public concerts, conducting and composing,
and by giving lessons and recitals to the gentry in their own
homes. Appointed in 1766 to the choice post of organist to
the fashionable Octagon Chapel, he set up in Bath. Bath
concentrated in one place the opportunities which had been
geographically scattered in the North, and Herschel
flourished there as a player, organist, composer, concert-
master and music teacher. By the early 1770's he could
harvest over £400 p.a. from musical performances and
teaching: the income of a respectable gentleman.[1]

For Bath was a good cradle for a career in the arts. Admittedly, it was provincial, and no provincial town could hold a candle to London. At the turn of the eighteenth century as many as forty percent of all townsdwellers in England lived in the capital. London had a population of close on 600,000; the next largest city was Norwich with 30,000 inhabitants; Bath had about 2,000.[2] As late as 1700 England possessed only one fashionable, polite, prestigious culture in the arts and sciences, the metropolitan, the home of the Court, of coffee-house culture, of the Royal Society. Outside London, culture merely glimmered with a few reflected beams; outside London, there seemed a wasteland of rusticity.

Oliver Goldsmith vividly recreated this scene in 1762, looking back to the beginning of the century:

> At this time, *London* was the only theatre in
> *England*, for pleasure, or intrigue. A
> spirit of gaming, had been introduced in the
> licentious age of *Charles* II and had by this
> time thriven surprizingly. Yet all its
> devastations were confined to *London* alone
> .... *Bath*, *Tunbridge*, *Scarborough* and other
> places of the same kind here, were then
> frequented only by such as really went for
> relief; the pleasures they afforded were
> merely rural, the company splenetic, rustic
> and vulgar.[3]

Bath was symptomatic of this provincial urban neglect. Francis Fleming was to write with some exaggeration:

> Bath, in the year 1670, was one of the
> poorest towns in England; so that four or
> five families residing here at one time
> rejoiced the inhabitants: the houses were
> very indifferent, there being only one
> that had a sash window .... There was
> neither ball-rooms, or places of amusement

.... Accommodations were but indifferent,
few houses were capable of receiving a
family of condition.[4]

Aspects of this imbalance of course persisted. Yet
many provincial towns, and Bath more than most, were trans-
formed during the Georgian century. They mushroomed in
size. Bristol for instance went up from about 20,000 in
1700 to about 80,000 in 1800; Manchester from about 10,000
to about 84,000; Bath from its 2,000 to 34,000. More or less
the whole of Bath was rebuilt - "Bath shoots out into new
crescents, circuses, squares every year", exclaimed Horace
Walpole.[5] Growth in size mirrored growth in wealth - the
product of agrarian prosperity, booming trade, particularly
with the expanding empire, and, especially later in the
century, industrialization. And, in consequence, opulent
townsmen wished to baptize their new-found wealth with
culture and style. Burghers in the localities did not want
to feel like *provincials*, rustics, hicks; they wanted to pass
as refined, polite, civilized, urbane.[6] To be precise, aided
by the new invention of the newspaper[7] (which told provincials what was going on at the nerve-centre of fashion), and
by the improvement to roads which turnpiking brought (the
journey from London to Bath was reduced from three days to
one), provincials clamoured to imitate London.[8] Corporations
and speculators built London-style theatres (often naming
them Drury Lane, or the Haymarket), and opened pleasure
gardens which they called Ranelagh or Vauxhall. In Bath the
architect John Wood the Elder laid out grounds "in imitation
of the Ring, in Hyde Park, near London".[9] Metropolitan
performers, such as the actress Mrs Siddons, came down to
give seasons. Performances of "Mr Shuter's London raree
show" were advertised in the *Bath Chronicle* for 16th
December, 1762. The first fashionable theatre in Bath was
floated by the London actor and impresario, Hippisley.[10]
Philip Astley, London's equestrian virtuoso, toured the West

country early in his career.

In short, converting yourself from rudeness to refinement meant making the mental journey from provincial to
metropolitan culture. In 1761 it was claimed that two
generations back the inhabitants of counties distant from
London had been "a species almost as different from those of
the metropolis as the natives of the Cape of Good Hope".
Now, at least the more "respectable" provincials had been
improved by the percolation of London styles and *mores*:

> the several great cities, and we might add
>
> many poor county towns, seem to be uni-
>
> versally inspired with an ambition of
>
> becoming the little *Londons* of the part of
>
> the kingdom wherein they are situated.[11]

No wonder a Newcastle address to the metropolitan rulers had
flattered:

> Our eyes are upon you; we ... imitate your
>
> fashions, good or evil, and from you we
>
> fetch and frame our customs. [12]

Of these towns which donned an elegant tone, Bath was
the most Londonized. "The Bath theatre", wrote the Rev.
John Nightingale in 1819, "is little inferior, in elegance
and attraction, to those of the metropolis".[13] Even the
Bath Penitentiary for Reformed Prostitutes was proud to
model itself on the London original - as presumably were the
prostitutes themselves. Some objected to this servile
mimicry. As Pierce Egan complained:

> And London fashions rattling down
>
> To make ye yet more overgrown,....
>
> In short, thou art so LONDONIZ'D
>
> So *over-built*, and *over-siz'd*,
>
> That, my old friend, I scarcely knew,
>
> Since last I said, dear BATH, adieu.[14]

But most inhabitants basked in the newly-achieved class and
elegance.

Bath won her eminence through being not just an
expanding provincial centre, close to opulent Bristol, but
by being a spa, a resort - in fact, in Horace Walpole's words,
the choicest of the "watering places that mimic a capital".[15]
That the waters had therapeutic properties had of course been
well-known at least since Roman times, and throughout Tudor
and Stuart times invalids came to take the waters - internally
and externally. But it was in Georgian times that
Bath became what Defoe puritanically called the
resort "of the sound rather than the sick". In the days
before the invention of the seaside, Bath was the nation's
leading resort, England's Las Vegas.[16] Visitors flooded in,
from gouty peers and politicians down to the newest *nouveaux
riches* tradespeople. They came above all for gambling, for
fashion, for the marriage market, for society, and some for
amours (though, despite Charles Wesley's dubbing Bath "the
headquarters of Satan", the town never became sexually
notorious like Charles II's Tunbridge Wells). "The goddess
of pleasure", it was said, "has selected this city as the place
of her principal residence". Amusement, not the Muses, was
Bath's business. The success of the Master of ceremonies,
Beau Nash, as Bath's Godfather lay in orchestrating the idle
into a genteel clockwork round. Tea-table frivolity and the
circulating library set the tone. Aside from architecture,
perhaps only in music did the demands of fashion stimulate
an inventive artistic tradition, deploying the talents of
Herschel, Linley, and then, *par excellence*, Rauzzini.

So William Herschel the budding musician made a wise
choice in coming to this vast pleasure dome. Yet he also
had longstanding mathematical, philosophical and optical
interests, and these began to claim a larger share in his
life from the early 1770's as he started buying optical
equipment and grinding lenses for telescopes, beginning to
sweep the skies with the aid of his sister Caroline. It is
just possible that he offered scientific tuition, but it is
more likely that he was scientifically isolated at Bath, for

all his scientific contacts in this early period were with men
outside Bath, e.g. Thomas Hornsby, Savilian Professor in
Oxford.   Herschel almost certainly was not famed in Bath for
his scientific bent, for his countryman the philosopher
Lichtenberg, visited the town in 1775 without even discovering
Herschel lived there:

> "Good heavens! Had I but known, when I spent
> some days in Bath in October 1775, that such
> a man was living there! Being no friend of
> tea-rooms and card-playing, I was very much
> bored there."[17]

But does this mean that Herschel's Bath was utterly
indifferent to science?   Barbeau, the greatest historian of
Georgian Bath, inclined to think so: "there was", he wrote,
"little desire for literary or scientific knowledge".[18]   But
this verdict is too austere.   Bath may not have been Athens,
but it was at least Corinth.   It was not a powerhouse of
profound researches, but there was a lively and widespread
taste for scientific knowledge.   From quite early in the
century, itinerant lecturers, armed with chemical apparatus
and orreries to explain the Newtonian heavens, had paid their
calls at Bath and Bristol.   William Whiston lectured at
Bristol in 1724.   James Ferguson - a typical Scot who had
taken the high road South to London and become a foremost
text-book popularizer of science - lectured in Bath and
Bristol in the 1760's and 70's - Herschel may have attended.
Benjamin Martin, James Arden, Henry Moyes, John Warltire and
others brought science down to Avon.[19]

Certainly, such men aimed to delight as much as to
instruct, and science became another mode of elite entertain-
ment.   As Goldsmith noted somewhat primly, "people of fashion,
when so disposed, attend lectures on the arts and sciences,
which are frequently taught there in a pretty superficial
manner, so as not to teize the understanding, while they
afford the imagination some amusement".[20]   Nevertheless, a

general buzz of interest existed.

And it was supplemented by knots of active participation. In 1777 the Bath and West Agricultural Society was founded and housed in Bath.[21] Local gentry flocked to join, willing to support its scientific experiments on soil, fertilizers and stockbreeding, since they thought that science in turn would support husbandry. It was to become a leader amongst Britain's agricultural societies, publishing its own *Transactions*.

In addition, Bath and its environs had a good sprinkling of enthusiasts for the natural history sciences, men such as Ralph Schomberg, William Watson,Jr, Thomas Haviland, Caleb Parry and John Walcott, collectors of plants and fossils, some of them keen to develop a systematic natural history of the West Country.[22] Bath has been claimed as the "cradle of British geology", for it was partly through the encouragement of the Bathonian Benjamin Richardson, and Joseph Townsend, the rector of nearby Pewsey, that William Smith (so-called "father of British stratigraphy") was launched on his geological career. Smith later repaid the debt by using his geological expertise to solve a major problem of seepage from the springs that supply Bath's waters. Not least, the Bath medical community – men such as William Oliver, William Falconer, William Moyses, Archibald Cleland and Charles Lucas – comprising probably the biggest concentration of physicians, surgeons, and quacks outside London – stimulated a certain level of scientific controversy, partly through the endless stream of acrimonious pamphlets they produced, disputing the chemical, physical and mineralogical content and therapeutic efficacy of the waters.[23]

These disparate and fluid groupings came together – albeit briefly – in the Bath Philosophical Society, founded in 1779.[24] This Society was remarkable perhaps less for what it accomplished than for its existence and since we have no minute books or published *Transactions* it is not very easy to

say just what it did.   Discounting the Lunar Society of
Birmingham, which was an informal club of friends, the Bath
Society was the first properly constituted provincial
scientific society founded in Georgian England, antedating
the Manchester Literary and Philosophical Society by two
years.[25] It came into existence on 28th December 1779
following a suggestion from the Quaker botanist, Thomas
Curtis, that there be set up a "Select literary society for
the purpose of discussing scientific and phylosophical
subjects and making experiments to illustrate them". His
Quaker friend, Edmund Rack, became the first secretary; there
were thirteen founding members - amongst whom were William
Herschel (described as "optical instrument maker and mathe-
matician") and his friend William Watson,Jr, F.R.S. - and
meetings were to be held weekly in winter and fortnightly in
summer.

The Society prospered for a while, hearing papers
across a wide range of the physical and natural history
sciences (though taking little interest in practical and
technological matters). But after the death of prominent
members, particularly the secretary Rack, and the removal of
Herschel to London, it lost its energies, and had collapsed
by about 1785. A second Bath Philosophical Society was
formed in 1799, but seems to have had a similarly brief
existence.[26]

Why did the precocious fire of organized science in
Bath burn itself out so quickly? It is partly because the
leading intelligentsia in Bath tended either to be visitors,
or at least footloose - men who, like Herschel, would move
away when opportunity offered (unlike the manufacturers such
as Wedgwood and Boulton who made up the core of the Lunar
Society, rooted in Midlands industries). Bath physicians
contributed surprisingly little. Whereas in Manchester and
Sheffield doctors invested their energies heavily into
science to win status for themselves as the guardians of

polite and rational values, in Bath physicians had social
position and status already; they directed their leisure into
poetry and letters.[27] This meant that the promoters of the
Bath Philosophical Society were in fact rather small beer.
It had no grandee patron, no elder statesman of science. Its
dynamic secretary, Edmund Rack, was a *petit bourgeois* Quaker,
a Uriah Heapish sycophant dedicated to getting on socially
while despising the very tinsel society he clawed to join, a
natural underling but commanding little authority or *éclat* of
his own - a man, in fact, much ridiculed in his day.[28]

Yet, for all its ups-and-downs, the Philosophical
Society may have been precisely the *stimulus* needed to launch
the obscure William Herschel onto a public stage and career
in science. Finding an outlet at last, Herschel proved an
irrepressible contributor to the Society. Within a month of
its foundation he gave his first paper, on Corallines; over
the next two years, he delivered thirty more, on subjects
ranging from metaphysics to natural history, electricity,
optics and of course astronomical observation. Some of these
were then communicated to the Royal Society and published in
Phil. trans.[29] The last was an "Account of a comet" -
Herschel's announcement of the discovery of Uranus read
March 1781. Was the Bath Philosophical Society midwife to
Herschel's astronomical revolution?

REFERENCES

1. *Cf.* C.A. Lubbock (ed.), *The Herschel chronicle* (Cambridge, 1933), *passim*; M.A. Hoskin, *William Herschel, pioneer of sidereal astronomy* (London, 1959).

2. A.E. Wrigley, "A simple model of London's importance in changing English society and economy, 1650-1750", *Past and present*, xxxvi (1967), 44-70; F.J. Fisher, "The development of London as a centre of conspicuous consumption in the sixteenth and seventeenth centuries", *Transactions of the Royal Historical Society*, xxx (1948), 37-50.

3. /Oliver Goldsmith/, *The life of Richard Nash of Bath* (London and Bath, 1762), 21. *Cf.* R. Lennard (ed.), *Englishmen at rest and play* (Oxford, 1931), 47f.

4. Quoted in V.J. Kite, *Libraries in Bath, 1618-1964* (thesis for fellowship of the Library Association, 1966), 12, from F. Fleming, *Life and adventures of Timothy Ginnadrake* (3 vols, Bath, 1771), iii, 15. For a necessary corrective, bringing out the genuine vitality of seventeenth century Bath, see P. Rowland James, *The baths of Bath in the sixteenth and early seventeenth centuries* (London, 1938). Nevertheless it is important to note how impressed were eighteenth century commentators with the rate, extent, and glory of Bath's current development. For introductions to the Georgian municipal growth of Bath see B. Little, *Bath portrait* (2nd ed., Bristol, 1968), and Sylvia MacIntyre, "Towns as health and pleasure resorts: Bath, Scarborough and Weymouth, 1770-1815" (D.Phil. thesis, Oxford University, 1973).

5. W.S. Lewis (ed.), *Horace Walpole's correspondence*, xi (New Haven, 1944), 288.

6. These beliefs were well articulated by John Wood the Elder, the Bath architect; for which see Ron Neale, "Society, belief, and the building of Bath, 1700-1793", in C.W. Chalklin and M.A. Havinden (eds), *Rural change and urban growth, 1500-1800* (London, 1974), 252-80. For Enlightenment respect for urban culture, see C.E. Schorske, "The idea of the city in European thought, Voltaire to Spengler", in O. Handlin and J. Burchard (eds), *The historian and the city* (Cambridge, Mass., 1966), 95-114.

7. G.A. Cranefield, *The development of the provincial newspaper, 1700-1760* (Oxford, 1926); R. McK. Wiles, *Freshest advices* (Columbus, Ohio, 1965).

8. See the suggestive remarks in S. Rothblatt, *Tradition and change in English liberal education* (London, 1976), ch.iv,

'London'.

9.   J. Wood, *An essay towards a description of Bath*, 2 vols.
      (London, 1749), 439-40; and more generally J.H. Plumb, *The
      commercialization of leisure in eighteenth century England*
      (Reading, 1973); J. Money, *Experience and identity,
      Birmingham and the West Midlands, 1760-1800* (Manchester, 1977);
      D. Read, *The English provinces. c.1760-1960* (London, 1964);
      P. Borsay, "The English urban renaissance: the development of
      provincial urban culture, c.1680-1760", *Social history*, vi
      (1977), 581-603; P. Clark and P. Slack, *English towns in
      transition 1500-1700* (Oxford, 1975). For a contemporary
      reflection, see "Present state of the manners, society, etc.,
      etc., of the metropolis of England", *Monthly magazine*, x
      (1800), 35-38.

10.   For Hippisley, see A. Barbeau, *Life and letters at Bath in the
      xviii century* (London, 1904), 65. For a good account of
      relations between London and provincial theatre see A. Hare,
      The Georgian theatre in Wessex (London, 1958).

11.   Quoted in Read, *op.cit.* (ref. 9), 18-19.

12.   Quoted in Clark and Slack, *op.cit.* (ref. 9), 156.

13.   Rev. J. Nightingale, *The beauties of England and Wales*, xii
      (London, 1813), 427.

14.   Pierce Egan, *Walks through Bath* (Bath, 1819), 58.

15.   Quoted in *The book of Bath*, written for the 93rd annual meet-
      ing of the British Medical Association (Bath, 1925), 79.
      Spas developed as a pioneer form of leisure resort, since
      they already had long existed as medical centres. W. Addison,
      English spas (London, 1951); D. Gadd, *Georgian summer: Bath in
      the eighteenth century* (London, 1970).

16.   See for instance, J.A.R. Pimlott, *The Englishman's holiday*
      (London, 1947); and J.A. Patmore, "The spa towns of Britain",
      in R.P. Beckinsale and J.M. Houston (eds), *Urbanization and
      its problems* (Oxford, 1968), 47-55.

17.   Quoted in M.L. Mare and W.H. Quarrell, *Lichtenberg's visits to
      England* (Oxford, 1938), 95.

18.   Barbeau, *op.cit.* (ref. 10), 111.

19.   For itinerant scientific lecturers see F.W. Gibbs, "Itinerant
      lecturers in natural philosophy", *Ambix*, vi (1960), 111-17;
      A.E. Musson and E. Robinson, *Science and technology in the
      Industrial Revolution* (Manchester, 1969), which *inter alia*
      charts the movement of science from London into the provinces.
      See also M. Rowbottom, "The teaching of experimental philo-
      sophy in England, 1700-1730", *Actes du XIe congrès inter-
      nationale d'histoire des sciences*, iv (Warsaw, 1968), 46-53.
      See E. Henderson, *Life of James Ferguson, F.R.S.* (Edinburgh,
      1867), 268, 278, 338, 407. Ferguson's *Astronomy* may have been
      the first astronomy text Herschel purchased. See Lubbock,
      op.cit. (ref. 1), 60. See also J.R. Millburn, *Benjamin Martin:
      author, instrument-maker, and 'country showman'* (Leyden,
      1976).

20.   /Goldsmith/, *op.cit.* (ref. 3), 46.

21.   K. Hudson, *The Bath and West* (Bradford-on-Avon, 1976).

22. H. Torrens, "Geological communication in the Bath area in the
     last half of the eighteenth century", in L.J. Jordanova and
     Roy S. Porter (eds), *Images of the earth* (Chalfont St. Giles,
     Bucks, 1979), 215-47.
23. J. Murch, *Bath physicians of former times* (Bath, 1882). For a
     taste of the balneological war, see W. Baylies, *Practical
     reflections on the uses and abuses of Bath waters* (London,
     1757); idem, *A narrative of facts, demonstrating the actual
     existence and true cause of that physical confederacy in Bath*
     (London, 1757); idem, *An historical account of the rise,
     progress and management of the General Hospital and Infirmary
     in the City of Bath* (Bath, 1758); for a bibliography of it see
     C.F. Mullett, "Public baths and health in England, 16th-18th
     century", *Supplements to the Bulletin of the history of
     medicine*, v (Baltimore, 1946), 1-85.
24. A.J. Turner, *Catalogue* to an exhibition, "Science and Music in
     eighteenth century Bath" (Bath, 1977); J. Hunter, *The
     connexion of Bath with the literature and science of England*
     (Bath, 1853), 81-83.
25. Roy Porter, "Science, provincial culture and public opinion in
     Enlightenment England", *The British journal for eighteenth
     century studies*, iii (1980), 20-46.
26. See Hunter, *op.cit.* (ref. 24). Of course, other fashionable
     centres like York and Norwich could not boast scientific
     societies at all. But my point is that such societies when
     founded in Birmingham, Manchester, Edinburgh, etc. went from
     strength to strength whereas in Bath they languished.
27. See above all A.W. Thackray, "Natural knowledge in cultural
     context: the Manchester model", *American historical review*,
     lxxix (1974), 672-709, p.685: "the medical profession as
     guardian of the polite virtues in an industrializing world".
     For an important exploration of the interface between the
     medical and literary worlds see G.S. Rousseau, "Matt Bramble
     and the sulphur controversy in the XVIIIth century", *Journal
     of the history of ideas*, xxviii (1967), 577-89.
28. Edmund Rack's "A disultory journal of events &c. at Bath"
     (Bath Reference Library R69/12675) for 1779 well expresses
     his own envy and disgust towards fashionable Bath society.
     For published expression of Rack's views and values see his
     Poems on several subjects (Bath and London, 1775), and his
     Mentor's letters (Bath and London, 1778), 23f. For a con-
     temporary characterization of Rack as a *petit bourgeois*
     social climber see /?P. Thicknesse/, *Edmund: an ecloque*
     (n.p., n.d.), 8:
              When from the land of Essex first I came,
              Propell'd by vanity and thirst of fame,
              Eager I strove in wild ambition's fits,
              To elbow in, and shine among the wits.
     See also /P. Thicknesse/, *A letter from Philip Thickskull,
     Esq., to Edmund Rack, a Quaker* (Bath, n.d.).
29. Herschel's papers to the Society are listed in J.L.E. Dreyer,
     The scientific papers of Sir William Herschel, 2 vols
     (London, 1912), i, pp.lxv-lxvi.

HERSCHEL'S SCIENTIFIC APPRENTICESHIP AND THE DISCOVERY OF URANUS

J.A. BENNETT
Whipple Science Museum
Cambridge CB2 3RH

In 1784 Jean-Dominique Cassini, who as Director of the Paris Observatory was one of the foremost professional astronomers of his day, wrote

> 'A discovery so unexpected could only have singular circumstances, for it was not due to an astronomer and the marvellous telescope....was not the work of an optician; it was Mr Herschel, an English musician, to whom we owe the knowledge of this seventh principal planet (Schaffer, 1981, 21).

Cassini later altered his account to describe Herschel as a German musician. Astronomers were generally taken aback and not a little confused by the emergence of this musician from relative obscurity. He was, it seemed, possessed of uncommon astronomical interests, unconventional methods and well-nigh unbelievable instruments. At the same time he was unfamiliar with the norms and conventions that governed communication within the established community. Yet by the time Cassini was writing it was clear that Herschel had to be taken seriously, for he had a single outstanding achievement to his credit - he had added a primary planet to the Solar System, while the other planets had all been known before the beginning of written astronomy.

What were the astronomers to make of this musician? What was his background, his training, his knowledge of astronomy? Although the circumstances of Herschel's early life are fairly well documented, these questions can scarcely be better answered today, and in this paper we will look specifically at his scientific apprenticeship. What was its content? How did it prepare him for the discovery? How did it shape his reaction to the opportunity the discovery presented?

We know that Herschel did not have a formal education in science, that he learnt his astronomy from the textbooks of his day in parallel with his own practical experiments and observations. The only clue we have to why he began is his own remark that a professional interest in musical theory led him to a general study of mathematics and in turn to a particular interest in optics and astronomy (Dreyer, 1912, xix). Robert Smith's textbook Harmonics, or the philosophy of musical sounds, according to this account, was naturally followed by his Compleat system of opticks in four books .... a popular, a mathematical and a philosophical treatise.

Given Herschel's character, the account is plausible enough. His sister Caroline once referred to the 'uncommon precipitancy which accompanied all his actions' (Lubbock, 1933, 67). She was referring to a characteristic physical enthusiasm, and physical energy and stamina were to be vital to his chosen astronomical career, but the same comment might be made of his intellectual character. He was ever alive to new interests and fresh poss- ibilities, and with great resourcefulness and single-mindedness he undertook programmes of study and research that were ambitious in the extreme. In the light of his other undertakings it is relatively easy to imagine Herschel coming to astronomy through reading widely in the mathematical sciences.

There are two notes of astronomical interests in his diary or 'memoranda' for 1766 (Dreyer, 1912, xix), but nothing of significance until 1773. In April 1773 Herschel bought a Hadley quadrant and a copy of William Emerson's textbook The elements of trigonometry (ibid., xxii, xxiv). An octant or Hadley quadrant was, of course, commonly used for finding latitude at sea, and seems an unlikely instrument for him to choose. By 1778 he was using it to check his clock by the method of equal altitudes (RAS MSS Herschel W2/1.1, 78). In May 1773 he bought what was, so far as we know, his first book on astronomy (Dreyer 1912, xxii, xxiv). This was James Ferguson's Astronomy explained upon Sir Isaac Newton's principles, and made easy to those who have not studied mathematics.

In the same month Herschel began to construct telescopes, but these were not the reflectors with which he is always associated.

He began in fact by buying object glasses of 4, 12, 15 and 30 feet
focal lengths and mounting them in tubes (Dreyer, 1912, xxiv). The
problems of managing long telescopes persuaded Herschel to turn to
reflectors, and by September he had a copy of Smith's Opticks and
was starting to grind and polish mirrors. It is interesting to see
that telescopes were playing an important part at the very beginning
of Herschel's astronomical interest. Their importance continued
throughout his career, and I will argue that it was his early
success in telescope building that largely determined Herschel's
eventual specialist interests in astronomy.

What other books did Herschel read during his apprenticeship?
In October 1773 he bought Emerson's The elements of optics (ibid.,
xxii), which like Smith was based firmly on Newtonian theory. It
was technical and fairly dull. Emerson dealt with instruments,
though in a less practical way than Smith, but unlike Smith did not
treat any astronomy. By 1776 Herschel had added to his library
Emerson's The principles of mechanics (ibid., xxv), which dealt with
both theoretical and practical mechanics, including the design of
machines, and so was useful in his practical work.

Also by 1776 we know that Herschel had Colin Maclaurin's
textbook on analytical geometry, A treatise on fluxions, and
probably a similar textbook by James Hodgson (ibid., xx). Herschel
later recorded that after a long day's work as a professional
musician and music teacher, he would use 'a few propositions in
Maclaurin's Fluxions' to, as he put it, 'unbend the mind' (Lubbock,
1933, 59).

By 1780 Herschel was familiar with Joseph Priestley's works
on light and on electricity, with John Keill's Introduction to the
true astronomy, based on his astronomy lectures at Oxford, and with
Lalande's Astronomie. (Dreyer, 1912, lxxii, lxxviii, xcvii, cv,
7, etc). In general we can say that Herschel was well served by
his informal education in astronomy. He had available textbooks
which were both sound and serious, and although popular, in the
sense that they assumed no knowledge of the subject, demanded
discipline and application if they were to be mastered.

There seems no doubt that Ferguson and Smith had the greatest

influence on Herschel. Smith gave him his basic grasp of casting,
grinding and polishing mirrors, constructing stands and applying
micrometers. Smith dealt also with observational astronomy by
running through all the phenomena that might be seen with a
telescope. Yet it is difficult not to feel that emotionally
Herschel had more sympathy, not with the Cambridge professor Robert
Smith, but with his fellow amateur and mechanic James Ferguson.

In his first chapter Ferguson presents his readers with a
dramatic and striking view of the universe, which in general terms
resembled the one Herschel adopted and attempted to work out in
detail. It was presented also with an enthusiasm which Herschel
would have appreciated. Traditional, academic, respectable,
professional astronomy in the eighteenth century was concerned
with the solar system, with the ramifications of Newtonian celestial
dynamics and with the navigational, horological and geographical
applications of technical precision astronomy. To all this the
fixed stars were little more than a convenient backdrop. Ferguson's
enthusiasm, however, encompassed the whole universe, which he
presents as an immense three-dimensional heavens, having many
planetary systems all populated by rational beings, so that the
solar system was only one example of its kind.

> 'What an agust! what an amazing conception, if human
> imagination can conceive it, does this give of the works of
> the Creator! Thousands of thousands of Suns, multiplied
> without end, and ranged all around us, at immense distances
> from each other, attended by ten thousand times ten thousand
> worlds, all in rapid motion, yet calm, regular, and
> harmonious, invariably keeping the paths prescribed them;
> and these worlds peopled with myriads of intelligent beings,
> formed for endless progression in perfection and felicity!
> (Ferguson, 1778, 6)

Herschel, of course, made the starry heavens and their three-
dimensional arrangement his particular domain. Ferguson's
enthusiasm was no doubt one reason for this, though we shall see
that Herschel's telescopes were a more immediate stimulus. He also
enthusiastically adopted the idea that the whole universe was

populated by rational beings, and this included the Moon and the
other planets in our system. Ferguson too imagined rational
creatures on the Moon, even though he had argued that there was no
atmosphere.

We find other links with Herschel in Ferguson's text. Ferguson
suggests that a star's brightness is a fair indication of relative
distance, and that the Sun is a typical star (ibid., 38) - two
assumptions that would be important for Herschel. He also thinks
that comets have a role to play in refuelling the Sun (ibid., 39)
and the stars (ibid., 355) - another idea that attracted Herschel
and one that derived from Newton (Schaffer, 1980, 97). Incident-
ally, Ferguson also thought that comets were inhabited by beings
in an especially privileged position to appreciate the wonders
of the heavens (Ferguson, 1778, 39).

The section of Ferguson's book that was perhaps most
significant for Herschel was his treatment of the nebulae, which
he calls either 'lucid spots' or 'cloudy stars'. The Orion nebula
he describes as 'the most remarkable of all the cloudy Stars',
and says that

> 'It looks like a gap in the sky, through which one might
> see (as it were) part of a much brighter region. Although
> most of these spaces [the 'nebulae'] are but a few minutes
> of a degree in breadth, yet, since they are among the fixed
> Stars, they must be spaces larger than what is occupied by
> our Solar System; and in which there seems to be a perpetual
> uninterrupted day among numberless Worlds, which no human
> art ever can discover (ibid., 353)

Smith had painted a similarly intriguing picture of the Orion
nebula (Smith, 1738 ii, 447-8), and both accounts derived from a
short paper in the Philosophical transactions published in 1716
(Phil.trans., 29, 1714-16, 392) and generally attributed to
Halley.

My selections from Ferguson probably do not give an accurate
impression of his book. Much of the text is fairly mundane and
technical, though not advanced, but it is interspersed with his
enthusiastic view of the subject as a whole.

Herschel's first year of serious astronomy is evidence in itself of his 'uncommon precipitancy'. Between May 1773 and March 1774 he assimilated the astronomy of the popular textbooks, mounted several long refractors before rejecting their use for good, arranged to have blanks cast in speculum metal, learnt the basic techniques of grinding and polishing the mirrors by hand, mounted one finished primary mirror in a Newtonian telescope, and began his 'Journal' of observations. All of this was done during what time he could spare from his extremely demanding musical life. It is difficult to know how representative are the entries in his very meagre diary, but on 8 November he recorded 'Attended 40 scholars this week. Public business as usual', and on 15 November, 'Attended 46 private scholars; nearly 8 per day' (Dreyer, 1912, xxii). Spare time during November was spent polishing. For January, when he was setting up the telescope, he writes, 'Gave 6,7 and 8 private lessons every day'. For March he writes, 'Nearly the same number of scholars. Astronomical journal begun'.

The telescope was, as I have said, a Newtonian of $5\frac{1}{2}$ feet focal length, and an aperture of perhaps $4\frac{1}{2}$ inches (RAS MSS W.5/12.1, 2; Dreyer, 1912, i, 109). This was a very respectable size for a maker with only a few month's experience. To give some idea of contemporary limitations on reflectors, James Short, the famous maker who had died in 1768, offered a six foot reflector of one foot aperture, but the price was 300 guineas (King, 1955, 86) - more than the annual salary of the Astronomer Royal.

On 1 March 1774 Herschel began his astronomical journal in a folio volume that now forms the first of a series of twelve preserved in the Archives of the Royal Astronomical Society. (RAS MSS Herschel W.2/1.1-12). It is not surprising, in view of what he had read in Smith and Ferguson, that his attention on his first evening's observing was directed to Saturn and the Orion nebula. He observed Saturn again on 2 March and the Orion nebula on 4th. Volume one of the Journal is a fascinating record of Herschel's first original work in astronomy, and of more immediate interest than the subsequent volumes, for in the early years he recorded, not only his observations, but also his ideas, his

speculations and his plans for the future. The entry for 4 March,
for example, on the very first page, shows that already questions of
central importance in his future research were beginning to take
shape:

> 'Saw the lucid Spot in Orions Sword, thro' a 5½ foot
> Reflector; its Shape was not as Dr. Smith has delineated in
> his Optics; tho' something resembling it....from this we may
> infer that there are undoubtedly changes among the fixt
> Stars, and perhaps from a careful observation of this Spot
> something might be concluded concerning the Nature of it.
> (RAS MSS Herschel W.2/1.1, 1)

During March and April of 1774 Herschel was mainly interested
in observing Saturn and its satellites, but he looked again at the
Orion nebula, and he also found his first double star:

> 'Observed the last but one in ursa Major's tail which is a
> double Star, and found when I magnify'd 211 times that it
> appeared very plainly to be double; being then separated
> nearly (as one might say) a couple of inches the lower being
> considerably larger than the other. (RAS MSS Herschel
> W.2/1.1, 5)

These few observations mark the end of Herschel's first burst
of astronomical activity, at least so far as we can discover from
surviving manuscripts. It had lasted for about a year and for some
two years from April 1774 only a few observations are recorded.
Some of these are of eclipses of Jupiter's satellites, and they
occasioned Herschel's earliest contact with the professional
astronomical community, for he exchanged letters on the subject of
these eclipses with the Radcliffe observer, Thomas Hornsby (RAS MSS
Herschel W.1/13. H.23, 24).

Just as we might imagine Herschel's enthusiasm to be waning,
a note in the Journal, which can be dated approximately to the
summer of 1774, gives an important clue to how his plans were
developing:

> 'If the nearest fix'd Star be 32 Billions of miles from our
> Sun, and of the same Bigness what angle will it subtend at
> the naked Eye and what must be the Magnifying power of a

Telescope to make it of any visible Diameter. (RAS MSS
Herschel W.2/1.1, 7)

He then devotes a few pages to this problem. Herschel had already
had the pleasure of separating double stars, and was clearly
beginning to wonder how far he would be able to go in improving his
telescopes and applying them to the stars. In particular, might
he be able to enhance his light-grasp and definition such that the
application of very high magnifications would reveal the true
apparent diameters of the stars? It is significant that the next
developments of any importance in Herschel's astronomy involved a
striking improvement in his telescopes.

The year 1776 marks the second surge in Herschel's activity.
We have little record of his work on their construction, but in
the space of three months he introduced three new telescopes and
dramatically increased his observing capability. We have seen
already that he had in mind an attempt on the true apparent
diameters of the stars, and for this he would need to increase not
only magnification but also light-grasp and quality of definition.

On 1 May 1776 he noted 'Observed Saturn with a New Reflector
Focus 7ft'. On 28 May he introduced a 10ft reflector,and on 13 July
he wrote 'I had a very good view of Saturn with a new reflector of
20ft Focus' (ibid., 13, 25). Telescopes of 7, 10 and 20 feet
focal lengths, whose apertures were respectively 6.2, 9 and 12
inches, served Herschel for the remainder of his time at Bath, and
he would have nothing larger for over seven years. Yet in the
meantime there were plenty of improvements to be made.

First the stands were continually being reviewed and underwent
frequent changes, aimed at more convenient management (for details
and references on the telescopes, Bennett, 1976). We know that
in 1778 Herschel arrived at the familiar design used for the 7ft and
10ft telescopes. It was a model of economy and convenience, for
the observer had every motion - coarse and fine vertical and fine
lateral - ready to hand while viewing at the Newtonian focus, and
in addition could observe in almost any altitude from a comfortable
standing position. The drawing by William Watson of Herschel's
own 7ft was made in 1783, and the design was used subsequently for

all the 'small' telescopes he made.

The mounting of the 20ft (the 'small 20ft'), however, was far
from convenient. For some time Herschel had to be content with
modifying the standard mount for a long refractor, which he must,
of course, have used himself in 1773. This consisted simply of a
single pole and a system of pulleys for raising one end of the
tube: essentially the same arrangement as had been used since long
refractors were introduced in the mid-seventeenth century. A long
reflector, however, was a novelty and Herschel, of course, needed
to position himself close to the Newtonian focus - hence the
simple expedient of a free-standing ladder. The result was somewhat
crude and ad hoc, though Herschel, again using strings and pullies,
managed to arrange fine vertical and lateral motions controlled
by the observer. The result was not only ad hoc but uncomfortable
for the observer, and very susceptable to wind disturbance. On one
night in April 1777 he gave up after, as he noted, 'the uneasy
posture and cold prevented farther Observ:' (RAS MSS Herschel W.2/
1.1, 42) but in general his perseverance with this instrument was
admirable. On later occasions he was even prepared to continue while
keeping his ink bottle in his hands, to prevent the contents freezing
(ibid., 5, 16).

While developments in stands represent one line of improvement,
figuring and polishing the mirrors was of more critical importance.
All of Herschel's early grinding and polishing was done by hand -
by moving the speculum metal mirror on top of the brass tool or the
pitch polisher - and the manipulative skills he acquired by long and
laborious experience were fundamental to his success as a telescope
maker. A serious programme of long and tedious experiments in
polishing began in earnest in March 1778, with Herschel trying all
sorts of materials and techniques and carefully recording the results
he achieved. Eventually the mirrors improved markedly in quality,
and in particular, on 14 November, recording a repolishing of one of
the primary mirrors for the 7ft, he says 'I used the divided
reducing stroke of the 170th experiment, and in a very short time
made it a most capital speculum' (RAS MSS Herschel W.5/12.1, 42).
It was when using this mirror that he later discovered Uranus.

But before this a third component in the link between the
heavens and the astronomer would have to be refined. In addition to
his stands and his mirrors Herschel would need to develop his own
sensitivity as an observer, and this he proceeded to do in a systematic way. 'Seeing is in some respect an art, which must be learnt',
he wrote to William Watson (RAS MSS Herschel W.1/1, 17-18), and his
learning technique was disciplined in a way not unlike his musical
training. We have seen that he was interested in applying the
maximum possible magnifying powers, and when he was able comfortably
to observe with a particular power, he would then purposefully apply
a higher one than he found easy to use (ibid., 27). When Herschel's
work eventually became known, the extravagant powers he claimed
sometimes to use (of up to 6,000) were generally doubted or frankly
disbelieved, and Herschel explained the ability he had acquired
by a musical comparison: 'To make a person see with such a power
is nearly the same as if I were asked to make him play one of Handel's
fugues upon the organ' (ibid., 17-18).

It is important to notice how early Herschel was equipped with
the best reflecting telescopes in existence. His telescopes and
his research programme always influenced each other, and it is
difficult to say at any time which was determining the character
of the other. Questions such as the nature of the Orion nebula and
the true apparent diameters of the stars were early stimuli to
building telescopes with large apertures, but Herschel's early
success in telescope making was, in its turn, an important
determinant in the kind of astronomy he would undertake.

We can take up the record of the Journal once again to
discover what Herschel was observing with his greatly enhanced
range of instruments. The 7ft we saw was first trained on Saturn
and on 28 May he writes:

> 'This evening I tryed a new ten foot reflector first on the
> Moon. with the Eye glass it had, it magnifyed 240 times very
> distinctly. The Moment I saw the Moon I was struck with the
> appearance of something I had never taken notice of before
> which I immediately took to be woods or large quantities of .
> growing Substances in the Moon.' (RAS MSS Herschel W.2/1.1, 13)

The next page or two are devoted to this exciting discovery and to
drawings of the lunar forests, with the conclusion that the
appearance of the so-called seas 'can be solved no other way than
by admitting....[them] to be Forests or some kind or other of Trees
or Plants.' Naturally the Moon was observed again a number of
times during 1776, as was Saturn. There is also one detailed
account of the Orion nebula.

The general impression of the Journal so far is that Herschel
had no very definite aims - no research programme that might
regulate the telescopes' use in a systematic way. The observation
pattern seems largely unstructured. The same is true of the year
1777, though he did begin to observe Mars, Venus and Jupiter, as well
as Saturn, and became interested in the variable star Mira Ceti.
This marks the beginning of work that would eventually form part
of the first three papers he presented to the Royal Society, two
in May 1780 and one in January 1781. They concerned Mira Ceti, the
rotation of planets and the heights of mountains on the Moon. This
last paper was published only after Herschel had agreed to remove
passages about the lunar inhabitants (Dreyer 1912, xc - xci).

In January 1778 the Journal grants us another glimpse into
the development of Herschel's thinking. In several fascinating
pages, full of implications for what was to come, he begins to
speculate about the three-dimensional arrangement of the stars,
supposing that fainter stars are more distant and that in general
stellar magnitude is a gauge of distance. He also develops the
ideas behind the method that had first been suggested by Galileo
for detecting stellar parallax. Without a measured parallax the
quantitative information basic to a study of the heavens in three
dimensions was not available, and the traditional fixed instruments
with graduated arcs had so far failed to find it. If, however,
micrometric measures were taken of close pairs of double stars, a
parallax might well be detected, especially if the members differed
greatly in brightness. Some such pairs would be optical doubles,
with one member much more distant than the other, and the distant
star could be regarded as a fixed point against which to measure
the motion of the other. One particular advantage of the method was

that no attention need be paid to the known disturbing influences
such as refraction, aberration, etc. Herschel writes:

> '... it is evident that nothing can be gained by this Method,
> except that we hereby reduce the annual parallax of a Star
> to a quantity that may be estimated by actual Inspection and
> is not liable to the accidents that render the Observation of
> so small an angle with any degree of precision next to
> impossible. Let us therefore examine to what degree of
> perfection a Telescope must be had in order to discover a
> Parrallax [sic] on the supposition that it is but .1".'
> (RAS MSS Herschel W.2/1.1, 50-51)

So again the question is seen in relation to perfecting the
telescopes, and of course it is significant that he had such fine
instruments, well suited to the task, before the plan was conceived.

There followed a few attempts to find suitable double stars,
but nothing systematic, and Herschel quickly resumed his planetary
observations. However on 5 March he returned to the double star he
had first seen four years previously:

> 'I directed my Telescope out of Curiosity to the double Star
> in the Bear and if I am not very much mistaken in the Eye
> piece I formerly used for this purpose the Stars were closer
> together this Evening than when I observed them last. This
> shall be farther examined and ascertained by proper
> experiment.' (ibid., 72-3)

But for 12 March we find:

> 'To my great disappointment I found the Stars in the tail of
> Ursa Major just as I saw them three months ago, at least not
> visibly different.' (ibid., 75)

On the same evening he found no change in the Orion nebula.

So the observations continue on the usual subjects through
1778 and 1779 - Saturn, Jupiter, sometimes Mars, very occasionally
the Orion nebula. In July 1779, however, we come upon another
speculative interjection, and find Herschel back with his old quest
for the true apparent stellar diameters. For a moment he feels that
the perfection of his telescopes has now achieved this illusive
goal. On 17 July he writes:

'The Evening being very fine my Telescope bore a power of
280 and convinced me that the Stars are of a Sensible
magnitude as I could see Arturus's Body very well defined.
Also the 2^d Star in the Bear's tail which is double showed
both the bodies very distinct. from this some consequences
may be drawn different from what has been comonly believed
heitherto [sic].' (**ibid.**, 107)

When Herschel's work became generally known, the fact that his
telescopes showed the stars 'round and well defined' did indeed
cause a considerable stir among astronomers (Lubbock, 1933, 90-102).
He himself drew out the consequences he had mentioned on the
following evening, 18 July:

'I continued my Observation in the Stars found the Stars in
the Bear, the pole Star, Altair & Star in the Crown all of
a visible but unequal magnitude.

Question. Suppose a Star of the first magnitude equal to
our Sun, to subtend an angle of 1" at the Naked eye what
is its distance and what will be the annual parallax of the
Orbit of the Earth. Let the Sun be 32' or 1920" and let the
Distance of the Sun be 100 Millions of Miles. then the
Distance of that Star must be, 192,000,000,000 Miles.

Next, what will be the parallax of the Earth's Orbit, answer
near 2 minutes, or rather 4' taking the whole diameter of it.
Next question is - Suppose the añual parallax of the Dia^r
of the Orb to be 2" what will be the Size of a Star to subtend
an angle of 1" at the naked Eye.

The distance in that case must be 20, 626, 400 Millions, and
to subtend ~1" at that Distance the real Dia^r must be 107
times as large as that of our Sun. From this, it should seem
that the Optic illusions represent the Dia^r of the Stars much
larger than they are. Or on the other hand, that there are
hopes that an annual parallax may be found.

The Method of two Stars must be resumed. their Situation
ought to be near the pole of the ecliptic, and the difference
of their Magnitudes as great as possible, also their distances
should not exceed 2" or 3".

> This Method has the superior excellence that none of the known
> causes such as aberation, Nutation, refraction, or any kind
> of Libration in the Earth's axis can effect it.' (ibid., 107-
> 108).

So, if Herschel is within reach of the true apparent stellar
diameters, either, on the one hand, the stars are very much larger
than the Sun, or, on the other, the annual parallax by the method
of double stars can be found. On 17 August, 1779 he began what he
called his 'second review of the heavens' (RAS MSS Herschel W.2/2.1)
with a very clear purpose. It was a systematic search for double
stars, employing the 7ft telescope and using a power of 227, and
resulted in a catalogue of 269 specimens presented to the Royal
Society.

From that point on the discovery of Uranus was assured.
Herschel was now equipped with very fine telescopes and a visual
sensitivity atuned to their use, and the sky was to be completely
reviewed. He later wrote:

> 'It has generally been supposed that it was a lucky accident
> that brought this new star to my view; this is an evident
> mistake. In the regular manner I examined every star of the
> heavens, not only of that magnitude but many far inferior,
> it was that night its turn to be discovered.... Had business
> prevented me that evening I must have found it the next, and
> the goodness of my telescope was such that I perceived its
> visible planetary disc as soon as I looked at it..' (Lubbock,
> 1933, 78-9)

The close of 1779 is noteworthy for Herschel's meeting with
William Watson and his introduction to the Bath Philosophical
Society. During 1780 he presented papers to the Society on a range
of scientific topics, and two minor astronomical papers were
communicated by Watson to the Royal Society of London.

The year 1781 began with plans for a really large telescope.
Herschel wrote in his polishing journal:

> 'Having long ago intended to make a very large reflector,
> as soon as I should find myself sufficiently acquainted with
> the method of constructing specula, I now began to project the

following instrument.' (RAS MSS Herschel w.5/12.1, 48)
This was to have been a telescope of 30ft focal length with a
primary mirror originally intended to be of 4ft diameter, and work
was begun on the mounting. In principle the mount was not unlike
that for the small 20ft, though an observing platform was to be
included within the support for the tube.

On Tuesday 13 March 1781 Herschel was proceeding with the
review in the search for double stars and so observing with
his 7ft telescope. He found what he noted as 'a curious either
Nebulous Star or perhaps a Comet' (RAS MSS Herschel W.2/1.2, 23).
Uranus had, of course, been observed before and mistaken for a star.
With Herschel's telescopes it was noticeably non-stellar and had
at last been recognized as worthy of individual attention. He found
that, like a planet, its apparent size increased in proportion to
the powers he applied to it. On Saturday 17 March Herschel noted:

'I looked for the Comet or Nebulous Star and found that it is
a Comet, for it has changed its place.' (ibid., 24)

The assumption that he had found a comet was the natural one
to make, and Herschel quickly informed Maskelyne at Greenwich,
Hornsby at Oxford and the Royal Society (Schaffer, 1981, 13).
Although he continued to observe his comet, a determination of its
orbit would require the kind of instruments that Herschel did not
possess. As he wrote to Watson:

'... my apparatus being but ill-adapted to such observations
as are necessary to settle the orbit of a Comet, which may be
much better done in a regular Observatory, I resign it to
abler hands (ibid.)

Meanwhile Herschel pursued his search for double stars and made plans
for casting the primary mirror of his new telescope - an instrument
designed for what was now his chosen specialism, sidereal
astronomy.

The story of the immediate consequences has often been told
and I will not repeat it in detail. The recognition of the planetary status of the new discovery came gradually. As early as 4
April Maskelyne wrote to Watson of Herschel's 'comet or new planet'
(RAS MSS Herschel W.1/13.M. 14) and on 23 April he wrote to Herschel:

'I am to acknowledge my obligations to you for the communication of your discovery of the present Comet, or planet, I
don't know which to call it. It is as likely to be a regular
planet moving in an orbit nearly circular round the sun as a
Comet moving in a very excentric ellipsis. I have not yet seen
any Coma or tail to it.' (ibid., M. 15)

On the other hand, as late as February 1782, Hornsby could write to
Herschel of his 'comet':

'It is the fashion I think now to call it a new star or planet,
but I cannot help thinking that it will prove to be a comet.'
(ibid., H. 29)

Part of the problem at the time was semantic. Comets are
in orbit round the Sun, and this body was certainly placed in a very
much larger orbit than the known planets. The question of the
precise status of the comet, Hornsby notwithstanding, had come to a
head around November 1781, when Herschel's discovery had been judged
sufficient qualification for the Royal Society's Copley Medal. In
replying to Sir Joseph Banks's letter informing him of the award,
Herschel responded to the question of his comet's status with his
usual enthusiasm for speculation:

'... a Body is now exposed to the attention of Philosophers,
which may prove to be either a new Planet or perhaps a star
that may partake both of the nature of Comets & Planets,
and be, as it were, a Link between the Cometary and Planetary
Systems, uniting them together by that admirable connection
already discover'd in so many other parts of the creation....'

and since it will at least be visible for many years,

'... we may probably become perfectly acquainted with its real
nature & thereby obtain a still more extended view of the
wonderful order that reigns throughout the whole solar and
sydereal System'(RAS MSS Herschel W.1/7).

As well as the status of his comet the astronomers were during
1781 concerned with other questions relating to Herschel: his claims
of very large magnifying powers, doubts and disputes over his
micrometer measurements on the comet, his seeing the stars round and
well defined. There was also much curiosity over the musician from

Bath. Maskelyne's letter of 23 April is full of questions about
Herschel's telescopes, their stands, their micrometers, whether he
casts his own mirrors, etc. Many questions had to be answered
before Herschel's new status in the astronomical community was
assured.

He too would learn from this period of probation, for in his
isolation he had been genuinely unaware of how uniquely well
equipped he had become. When he eventually visited Greenwich
and compared his 7ft with the telescopes available to Maskelyne,
he realized the exciting possibilities that lay ahead. 'Let me but
get at it again!' he wrote to his sister Caroline, 'I will make such
telescopes and see such things...' (Lubbock, 1933, 116).

By the summer of 1782 the queries surrounding Herschel's comet
and his methods and apparatus had been sufficiently clarified for
his friends to lobby successfully for a permanent astronomical
position for him. Granted a stipend by George III, he moved from
Bath to near Windsor at the beginning of August.

The popular view of the significance of the discovery has
always been a romantic one, and justifiably so: an obscure amateur,
observing in his back garden with a home-made telescope discovers
the first primary planet to be found since the dawn of history. Yet
the more important significance for the development of astronomy was
that the discovery gave Herschel his opportunity: it gave him a
position in the scientific community and the chance to devote
himself to astronomy. As an outsider, unrestricted by the established
pattern of the professional astronomy of his day, he chose a novel
theoretical domain, for his subsequent planetary work was always
secondary to his work on what he called 'the construction of the
heavens'.

In this sense it was fortunate that the discovery fell to
Herschel, for who else possessed the energy and vision - the sheer
audacity - to undertake a programme that required him to build huge
telescopes for penetrating deep into space and to speculate in a
wholly new theoretical domain that embraced the entire universe?

We began with the comments of the Director of the Paris
Observatory. On 8 August 1782 Herschel had only just moved to

Windsor when the Astronomer Royal, Nevil Maskelyne, wrote him a
letter which seems to represent an official recognition of his
position in the astronomical community and an official sanction on
the new career he was about to begin:

> 'Astronomy and Mechanics are equally indebted to you for what
> you have done; the first [sic] for your shewing to artists
> to what degree of perfection telescopes may be wrought; and
> the latter [sic] for your discovering to Astronomers a number of
> hitherto hidden wonders in the heavens, which could not be
> explored before for want of telescopes equal to yours; and
> they are both likely to receive equal improvement from it
> in the construction of better telescopes, and in the appli-
> cation that may be made of them to the heavens for repeating
> and extending your observations. I hope you will do the
> astronomical world the faver to give a name to your planet,
> which is entirely your own, & which we are so much obliged
> to you for the discovery of.' (RAS MSS Herschel W.1/13 M 20)

Maskelyne could scarcely have guessed where it would all lead. With
Herschel's move from Bath the immediate consequences of the
discovery were complete. The far-reaching consequences had barely
begun.

REFERENCES

J.A. Bennett, '"On the power of penetrating into space": the
    telescopes of William Herschel', Journal for the History of
    astronomy, vii (1976), 75-108.
J.A. Bennett, 'Catalogue of the archives and manuscripts of the
    Royal Astronomical Society', Memoirs of the Royal Astronomical
    Society, lxxxv (1978), 1-90.
J.L.E. Dreyer, ed., The scientific papers of Sir William Herschel,
    (London, 1912).
J. Ferguson, Astronomy explained upon Sir Isaac Newton's principles,
    (London, 1778).
H.C. King, The history of the telescope, (London, 1955).
C.A. Lubbock, ed., The Herschel chronicle, (Cambridge, 1933).
S. Schaffer, '"The great laboratories of the universe": William
    Herschel on matter theory and planetary life', Journal for
    the history of astronomy, xi (1980), 81-111.
S. Schaffer, 'Uranus and the establishment of Herschel's astronomy',
    Journal for the history of astronomy, xii (1981), 11-26.
R. Smith, A compleat system of opticks in four books, (London, 1738).

NOTES

William Watson's drawings of the 7ft and small 20ft telescopes are reproduced in Bennett 1976.

I am grateful to the Council of the Royal Astronomical Society for permission to quote from manuscripts in their possession.

HERSCHEL AND THE CONSTRUCTION OF THE HEAVENS

MICHAEL HOSKIN
Whipple Science Museum
Cambridge CB2 3RH.

The history of astronomy knows great telescope builders, great observers, and great theorists; but only William Herschel falls indisputably into all three categories. When he became a professional astronomer in the summer of 1782, he had already demonstrated his skills in the construction of big reflectors: the mounting of his 20ft still left much to be desired, but otherwise his mirrors, his eyepieces, and his mountings combined to give him a head start over any other astronomer in the examination of distant and therefore faint objects. And in little over a year after his arrival near Windsor Castle where he was to be available on occasion to show the heavens to the Royal Family, he completed one of the great telescopes of all time: his 'large' 20ft reflector, with mirrors of 18 inches diameter, and soon to be equipped with a stable ladder-type mounting so that telescope and observing platform could be rotated together by a single workman. This mounting he further improved and refined in the years to come, and meanwhile the telescope was his favourite instrument and in constant use during his twenty-year systematic search of the sky for nebulae. In his extreme old age it was refurbished under his direction by his son John, who resurveyed his father's nebulae and then took the telescope to the Cape of Good Hope to extend the coverage to the southern skies. John's General Catalogue of Nebulae and Clusters of Stars (1864) led to the New General Catalogue that astronomers use today.

Herschel supplemented his pension by constructing telescopes for sale, so that he was indeed a professional telescope maker. For himself he built with help from the King a monster reflector of 40ft focal length, with mirrors 4ft in diameter and weighing up to a ton. The mounting was essentially a scaled-up version

of that of the large 20ft; but it proved cumbersome in use, and the
modified alloy of the mirrors led to rapid tarnishing: Herschel had
exceeded the limits of his technology. He did however complete
an excellent reflector for the King of Spain, with 2ft mirrors and
focal length of 25ft. Until its destruction by Napoleonic troops,
this was the finest telescope of its kind; but, like almost all
the telescopes Herschel made for sale, little use was made of it
by its owners.

Herschel's skill as an observer had been demonstrated by
his discovery of Uranus. This skill was allied to an heroic
endurance that astonished visitors to his home. One of them wrote:

> ... I went to bed about one o'clock, and up to that time
> he had found that night four or five new nebulae. The
> thermometer in the garden stood at 13° Fahrenheit; but in
> spite of this, Herschel observes the whole night through,
> except that he stops every three or four hours and goes into
> the room for a few moments. For some years Herschel has
> observed the heavens every hour when the weather is clear,
> and this always in the open air, because he says that the
> telescope only performs well when it is at the same
> temperature as the air. He protects himself against the
> weather by putting on more clothing (Lubbock, The Herschel
> Chronicle, Cambridge, 1933, 138).

Astronomers not privileged to see him at work would find in
Philosophical Transactions the fruits of his labours: two
catalogues of double stars, three catalogues of nebulae and
clusters adding two-and-a-half thousand to the hundred or so
already known, lists of stars in diminishing order of apparent
brightness designed to facilitate the detection of variables --
the outcome of observational campaigns of unprecedented persistence.

But they were also the signs of theoretical interests
hitherto unknown in astronomy. Herschel was creating a new
astronomy, an astronomy that was part of natural history, as the
factual underpinning of a speculative cosmology. In the investigation of the solar system (to which most contemporary astronomers
were committed), each planet and each satellite had its own name,

each comet appeared in a given year and had a recognisable
individuality. Herschel, by contrast, counted stars and drew
inferences from the numbers alone. And he was a natural historian
collecting innumerable specimens, double stars by the hundred,
nebulae by the thousand. He described and classified, first
according to superficial appearances, later into true species.
To teach the life-cycle of a nebulae he paraded before his readers
nebulae that he declared were young, middle-aged, and old, after
the manner of a botanist pointing out trees at different stages
of their growth. Such methods were totally alien to astronomy;
they were importations from natural history, a truly new astronomy.

This fact-gathering, so orthodox in other fields of scientific
enquiry, might have been painlessly assimilated into astronomy if
Herschel had been of the cautious temperament of most great
observers. Instead, with a frankness seldom equalled in science,
he announced in print his intention to speculate too much rather
than too little:

> If we indulge a fanciful imagination and build worlds of
> our own, we must not wonder at our going wide from the
> path of truth and nature; but these will vanish like the
> Cartesian vortices that soon gave way when better theories
> were offered. On the other hand, if we add observation to
> observation, without attempting to draw not only certain
> conclusions, but also conjectural views from them, we offend
> against the very end for which only observations ought to
> be made. I will endeavour to keep a proper medium; but if
> I should deviate from that, I could wish not to fall into
> the latter error (Phil. Trans., lxxv, 1785, 213).

As usual, Herschel was as good as his word. 'A knowledge
of the construction of the heavens', he wrote in Philosophical
Transactions in 1811 (p. 269), 'has always been the ultimate
object of my observations', and from his early days as a
professional astronomer to the end of his long life, he published
a succession of massive papers on the large-scale structure of the
universe.

His contemporaries did not know what to make of it. His

stubborn self-confidence verging on arrogance alienated some of them, while even his most loyal supporters were at times baffled and ill-at-ease. For his papers on the physical description of the solar system he received due credit; but these were mere asides from his central concern with the construction of the heavens, and here his contemporaries could not follow him.

They could not follow him because he alone possessed huge cosmological telescopes designed to reach far out into space. When Herschel described a nebula, other astronomers could either believe him or desbelieve him; they could not do what astronomers can normally do, and that is to look for themselves and judge whether what is claimed is in fact true.   And even if they believed his descriptions, still the theoretical questions to which these observations were directed had never been part of astronomy, and his methods were often alien importations from natural history. The accepted conventions for the reobservation and critical assessment of published science by fellow professionals, so fundamental in any scientific community, simply did not exist as far as Herschel's life-work was concerned. It was therefore impossible for the novel concepts and methods he created to be assimilated into astronomy in his own lifetime, and they were not so assimilated. He was honoured in his own time as a pioneer telescope builder, a dedicated observer, and the maker of several important discoveries in the solar system. His 'construction' papers, however sceptically received, were published in Philosophical Transactions and so were widely available and part of the permanent and accessible record of the science of his age. Future generations of astronomers would inherit the questions he asked, his methods, even his terminology like 'planetary nebula', 'asteroid', and 'solar apex'; and we today can look back on his achievements conscious of the promise they held for the future -- for Herschel virtually created stellar astronomy and scientific cosmology. But as we now survey briefly his major contributions to the study of 'the construction of the heavens', we must not be deceived into thinking these were appreciated by his contemporaries.

DOUBLE STARS

As we saw in the preceding article, Herschel's first major task
in astronomy was to collect double stars in the hope that some of
them would be formed of a nearby star and a second star so distant
as to be virtually a fixed point provided by nature for the
convenient measurement of the annual parallax of the nearby star.
He published catalogues of double stars in Philosophical Transactions
in 1782 and 1785, and a third as a token paper in 1821 in the
Memoirs of the newly-founded [Royal] Astronomical Society; but
he never seriously attempted to detect annual parallax. Instead,
like James Bradley with the discovery of the aberration of light,
Herschel was to earn an unexpected reward for his efforts.

Newton had declared gravity to be a universal law of nature,
but outside the solar system he had no evidence for this. John
Michell (Phil. Trans., lvii (1767), 234) had pointed out that double
and multiple stars were many times more frequent than one would
expect on the hypothesis that all were chance (line-of-sight)
configurations, and that most must be true physical associations.
Michell's paper was not known to Herschel when he began his search
for double stars, and in 1784 (Phil. Trans., lxxiv, 35) Michell
published a second paper in which he expressly warned that most
of Herschel's doubles would prove to be binary stars and so be
useless for the detection of annual parallax. That Michell was
right became clear when Herschel's twenty years of sweeping
for nebulae came to an end and he had time to re-examine some
of his double stars. The components of several, he announced
(Phil. Trans., 1803, 339; 1804, 353), had altered position in a way
that showed they were in orbit around each other. Of these the best
documented was Castor, for in about 1759 James Bradley had remarked
to Nevil Maskelyne that the line joining the components was parallel
to the line joining Castor to Pollux. This extension of the time-
span of Herschel's own observations enabled him to give the
double star a period of rotation of 342 years (modern value: 306).

Herschel had little doubt that the force binding the
components together was gravity, but observational proof of this
would not be available in his lifetime.

THE MOTION OF THE SOLAR SYSTEM

Until 1718 all the evidence had indicated that the stars were
indeed 'fixed' and motionless, for they were without exception in
the same positions relative to each other as recorded in Antiquity.
In that year Edmond Halley announced (Phil. Trans., xxx (1717-19),
736) that Aldebaran, Sirius and Arcturus had altered position
in latitude since the time of Ptolemy. But Bradley's subsequent
discovery of aberration revealed a major source of error in modern
star catalogues, and it was not until 1760 that proper motions of
stars were known in any quantity. In a lecture that year (Opera
inedita, 1775, VI: De motu fixarum proprio), Tobias Mayer listed
the changes in right ascension and declination of some eighty
stars whose modern positions he compared with those recorded half-
a-century earlier. Of these changes he considered fifteen or twenty
large enough to be due to true proper motions rather than instrumental
errors. Mayer also explained the pattern of proper motions that
would be generated if the solar system were moving towards some region
of the sky: 'all the stars which appear in that region would seem
to be gradually separating from each other one by one, and those
which are in the opposite part of the sky would seem to be joining
up...' (trans. E.G. Forbes, London, 1971, 112). No such pattern,
he declared, was present in the data he had given.

It was an uncharacteristic problem for Herschel to tackle,
for his telescopes and his observing programmes were irrelevant.
Instead, he would be working at his desk on data published by
other astronomers, though when he first tackled the problem in
the winter of 1782/83 he did not own a copy of Mayer's lecture.
Instead, he possessed Maskelyne's Astronomical Observations Made at
the Royal Observatory... 1765-1774 (London, 1776), which gave a
brief list of proper motions (seven stars in right ascension, and
Sirius and Arcturus in declination), and the supplementary vol. iv
of the second edition of Lalande's Astronomie (Paris, 1781) which
carried a list of the most convincing of Mayer's proper motions.

Very conveniently, Maskelyne's seven motions in R.A. were
all consistent in direction with the motion of the solar system
towards an 'apex' with R.A. between about 14^h and 19^h. As to

Lalande's stars, some were included because of their large components
of motion in declination, although their listed components in R.A.
were small and probably spurious. With typical insensitivity
Herschel took all these components in R.A. at face value, even the
negligible change of 3" in fifty years recorded for Aldebaran.
The majority of these components were again consistent in direction
with an apex between about 14^h and 19^h. Only one star lay well
inside this range or its opposite: Aldebaran; and by taking the
direction of Aldebaran's component as established (and refusing
to recognise it for the negligible quantity it was), Herschel
managed to halve the range of R.A. within which the apex of the
solar motion had to lie if the listed exceptions were to be
reduced to a minimum. The mathematically more difficult problem
of declination he tackled essentially by inspection, and so arrived
at a proposed apex near λ Herculis, astonishingly close to the
best modern positions. But while it is true that the data available
to Herschel suggested a region of sky where the apex might lie,
the proximity of his apex to modern positions owes much to his use
-- or rather misuse -- of the figures for Aldebaran.

Never one to leave well alone, Herschel returned to the
problem in papers published in Philosophical Transactions in 1805
and 1806. This time he tried to derive not only the direction of
the solar motion but also its velocity; and for this he needed first
to establish the velocities of stars relative to the solar system.
Nothing of course could be known of the line-of-sight components.
The transverse components had to be derived from the proper motions
(that is, angular velocities) multiplied by the distances. To
derive distances he had to assume that all stars are intrinsically
of the same luminosity, so that distances are related to apparent
magnitudes. As in fact stars vary enormously in luminosity,
Herschel became trapped in a tangle of argumentation. In particular,
he concluded that bright (and supposedly near) stars with no known
proper motions and apparently at rest must in fact be keeping pace
with the Sun, so that Herschel actually ended up with more motions
than he started with -- in contrast to the earlier investigation
whose purpose was to show that many observed proper motions were

merely optical effects of the solar motion. But good came out of
ill, for this led Herschel to discuss the star cluster of which the
Sun is part, and which he supposed to be formed of stars moving
through space with the Sun.

THE MILKY WAY

In the early years of the second half of the century, three
speculators -- Thomas Wright of Durham, Immanuel Kant, and J.H.
Lambert -- had each suggested that the Milky Way is the optical
effect of our immersion in a layer of stars. We do not know if any
hint of this reached Herschel; in any case, he determined to take
the further step of charting the outline of the Galaxy, and he
realised he could do this if he allowed himself two assumptions.
The first was that his telescope (the 'large' 20ft) could penetrate
to the borders of the system in every direction -- for otherwise
his enterprise was hopeless. The second, and more interesting,
was that within the Milky Way system, space is uniformly stocked
with stars, so that when we see more stars in a given direction,
this is because the system extends further in that direction (rather
than because the stars are more clustered).

Armed with these assumptions, Herschel systematically
counted the numbers of stars around a great circle of the sky (which
was all he could spare time for). He converted these numbers into
(relative) distances and published the resulting cross-section of
the Galaxy (Phil. Trans., lxxv (1785), 213). It was a dramatic
gesture that virtually created the method of stellar statistics.
But as time went on Herschel lost confidence in both his assumptions.
His 40ft telescope brought many more stars into view, so that the
20ft had not after all penetrated to the borders in every direction:
indeed, appearances suggested that even with the 40ft the Milky Way
was in some directions 'fathomless'. Further, as his sweeps for
nebulae and star clusters continued, so did Herschel's experience of
how widespread is the phenomenon of star-clustering, and he admitted
(Phil. Trans., 1817, 302) that higher star counts indicated
greater clustering as much as greater distance to the border. But
if his assumptions were erroneous, his technique of stellar

statistics was to become a basic tool in stellar astronomy.

NEBULAE

As we saw in the preceding article, on the very first page of
his first Journal, which he wrote in March 1774, Herschel commented
that the Orion Nebula was now different in shape from the sketch
in Robert Smith's Opticks 'and perhaps from a careful observation of
this Spot something might be concluded concerning the Nature of it'.
It is astonishing to find this musician for whom in 1774 astronomy
was still a recently acquired hobby putting his finger so quickly
on the crucial question; for if a nebula had altered shape in only
a few decades, then it could not be a huge star system disguised
by distance, but must be a small, nearby object and non-stellar in
nature. Only the Orion Nebula had been sketched in the seventeenth
century, and it was this sketch (by Huygens) that Herschel was
comparing with what was now to be seen.

Herschel that March sketched the Orion Nebula himself and
again in 1776 and 1778, and satisfied himself that it did indeed
change shape. But although his ('small') 20ft was one of the best
telescopes in existence for the examination of nebulae, it was not
until after his move to the Thames Valley that he began to examine
nebulae in quantity. He found on examination that 'most' of the
nebulae in a catalogue by Charles Messier 'yielded to the force of
my light and power, and were resolved into stars ' (Phil. Trans.,
lxxiv, 1784, 437). In the autumn of 1783, on the completion of his
'large' 20ft, he set about systematically 'sweeping' the sky for
nebulae. Some were clearly 'resolved' into stars by his powerful
telescope, but others remained milky in appearance. How to
distinguish star systems disguised by distance from 'true' nebulae
(if such there were)? Herschel satisfied himself that a star
cluster disguised by distance presented the appearance of 'mottled'
or uneven nebulosity that he therefore termed 'resolvable', while
true nebulae had the even appearance of 'milky nebulosity'.

As a natural historian who was collecting hundreds of nebulae,
Herschel found himself compelled to classify them, at first by
mere appearances but later, he hoped, into a natural classification.

As he reflected on his specimens, he realised that star clusters suggested the presence of a clustering force (gravitational attraction?), and that a widely scattered cluster might be expected to develop in time into a more condensed cluster. To this extent it would be proper to describe a scattered cluster as youthful and a condensed cluster as, by comparison, aged -- truly novel concepts in astronomy.

Meanwhile, Herschel was beginning to publish both catalogues of nebulae and theoretical papers on 'the construction of the heavens'. On 22 June 1784, just five days after his first 'construction' paper was read to the Royal Society, Herschel came across the Omega Nebula (M17), followed next month by the Dumb-bell Nebula (M27); and in both he found the two nebulosities, milky and resolvable, coexisting side by side. Since the resolvable nebulosity was, he believed, composed of great numbers of distant stars, might not the milky nebulosity be composed of great numbers of very distant stars? -- in which case he had been wrong in the past to equate milky with 'true' nebulosity, for he now realised that all the appearances could be accounted for in terms of star systems alone. All, that is, except for the observed changes in the Orion Nebula, which he proceeded to disregard.

In his second 'construction' paper (Phil. Trans., lxxv, 1785, 213), Herschel shows that many of the nebulae (that is, star clusters) he had observed could have come about through the action of gravitational attraction on an initial, widely scattered distribution of stars. And he explains that some extensive yet milky nebulae must be huge star systems that 'may well outvie our milky-way in grandeur'. Yet what was the ultimate fate of a star system? Were there repulsive forces to ward off gravitational collapse? If collapse took place, what then happened? And what role in the universe was played by the mysterious 'planetary nebulae' that Herschel had discovered and which looked like planets but might be nebulae or might be a previously-unknown kind of heavenly body?

On 13 November 1790, during his regular sweeping, Herschel came across NGC1514, which is in fact a planetary nebula of unusually large apparent diameter with a prominent central star. Herschel admitted that the star must be connected with the encircling nebulosity, and in a dramatic shift of position he accepted that the star was condensing out of what was therefore true nebulosity. Planetary nebulae, and 'nebulous stars' as NGC1514 was classified, now became stages intermediate between the nebulous phase and the stellar phase of the life history of a nebula/star-cluster; and in papers written towards the end of his career (Phil. Trans., 1811, 269; 1814, 248), Herschel selected from his catalogues specimens at every stage in their life-history, claiming thereby to be in effect permitting his readers to witness a nebula develop from infancy as scattered nebulosity to its old age as a tightly-packed globular star cluster.

In other papers Herschel permitted himself wide-ranging speculations on related topics. Where did the nebulosity come from? Perhaps it was light itself, emitted by stars and collected into clouds by the mutual attraction of the light particles. A comet might be a small cloud of nebulosity pulled in by the attraction of the Sun and giving some of its substance to replenish the Sun as it passes through perihelion. Other material in the comet may be consolidated by the heat of the Sun, and after many such passages the head of the comet may become planet-sized. Stars themselves are in reality only planets with a luminous outer atmosphere.

And so Herschel piled speculation on speculation. His contemporaries were often baffled and sometimes openly hostile. He was admitted to be a telescope builder on an heroic scale, an observer of exemplary dedication, and the author of many disoveries within the solar system. But the general reaction to his theories of the construction of the heavens is expressed in the 1820 'Address of the Astronomical Society of London, explanatory of their views and objects':

> Beyond the limits however of our own system, all at present
> is obscurity. Some vast and general views of the construction
> of the heavens, and the laws which may regulate the formation
> and motions of sidereal systems, have, it is true, been struck
> out; but ... they remain to be supported or refuted by the

slow accumulation of a mass of facts...

BIBLIOGRAPHY

The Scientific Papers of Sir William Herschel, ed. by J.L.E. Dreyer
    (2 vols, London, 1912).

Bennett, J.A., '"On the power of penetrating into space": the
    telescopes of William Herschel', Journal for the History of
    Astronomy, vii (1976), 75-108.

Hoskin, Michael, Stellar Astronomy: Historical Studies (Chalfont
    St Giles, 1981).

Hoskin, Michael, William Herschel and the Construction of the
    Heavens (London, 1963; includes reprints of Herschel's
    major papers).

Hoskin, Michael, 'William Herschel's early investigations of
    nebulae: a reassessment', Journal for the History of
    Astronomy, x (1979), 165-76.

Hoskin, Michael, 'Herschel's determination of the solar apex',
    Journal for the History of Astronomy, xi (1980), 153-63.

Schaffer, Simon, 'Uranus and the establishment of Herschel's
    astronomy', Journal for the History of Astronomy, xii (1981),
    11-26.

Schaffer, Simon, '"The great laboratories of the universe":
    William Herschel on matter theory and planetary life',
    Journal for the History of Astronomy, xi (1980), 81-111.

Schaffer, Simon, 'Herschel in Bedlam: natural history and stellar
    astronomy', The British Journal for the History of Science,
    xiii (1980), 211-39.

THE PRE-DISCOVERY OBSERVATIONS OF URANUS

Eric G. Forbes
University of Edinburgh

In his *Astronomisches Jahrbuch* for 1784, Johann Elert Bode[1]
summarises the scanty information that had reached Berlin concerning the recent discovery on 13 March 1781 of a new heavenly
body by a still-unidentified observer in England.[2] Its easterly
progress through the Milky Way during the interim six months had
been parallel to the ecliptic, and thus entirely consistent with
the view - hitherto based only upon its brightness and clearly-
defined disc - that it was a planet and not a comet. Bode therefore asks why this sixth-magnitude object had not been previously
detected, and raises the question of whether it had in fact been
observed by earlier astronomers but misidentified as a star.[3] He
himself had already scanned the star-catalogues of Tycho Brahe,[4]
Johann Hevelius,[5] John Flamsteed,[6] and Tobias Mayer;[7] and had come
to suspect that a missing sixth magnitude star in the constellation
Capricorn, observed by Tycho on 20 November 1589, might have been
the planet. A second possibility which still required investigation was Mayer's star No. 964, observed in Aquarius on 25 September
1756.[8]

In order to obtain more precise data on this later and much
more reliable observation, Bode wrote to Abraham Gotthelf Kaestner[9]
who, as Director of the Göttingen Observatory, had access to the
manuscript papers deposited there by Mayer's widow in 1763 in
return for a modest payment from the Hanoverian government.[10]
Kaestner placed this matter in the capable hands of his colleague
Georg Christoph Lichtenberg, the editor of a volume of *Opera*
inedita incorporating Mayer's zodiacal star catalogue.[11]
Lichtenberg duly supplied the equatorial co-ordinates for the star
in question and the precise date and time of observation.[12] With

the aid of Anders Lexell's elements for the new planet, Bode[13] calculated its longitude and latitude for that date and time and compared these co-ordinates with those of 'star' No. 964.   The celestial latitudes agreed completely, while the difference of almost 8° in longitude was attributed by Bode to uncertainties in the values adopted for the orbital eccentricity and mean motion. His conclusion that the identification was "highly probable" was supported soon afterwards by Herschel himself[14] and by Pierre Méchain.[15]   Using Mayer's co-ordinates to calculate the helio-centric longitude of Uranus and comparing it with that derived for Tycho's star 27 Capricorni, which he also identified provisionally with that planet, Bode calculated a periodic time of 80 years 8 months - in close agreement with the figure derived from Uranus's observed motion over the 2½-years since its discovery.[16]   Since, however, there was reason to believe that Tycho's catalogued values contained copying or printing errors, this result was not regarded as conclusive; and subsequent calculations were soon to cause Bode to abandon this supposition.[17]

Consequently, in March 1784, Bode began to consider the further possibilities of the missing stars 8 and 34 Tauri recorded in the second volume of Flamsteed's catalogue being Uranus.[18]   The possibility of 8 Tauri being Uranus was soon to be ruled out, however, by Pierre Charles Le Monnier's identification of it with a star that Flamsteed had observed on 29 September and 1 October 1704.[19]   According to Méchain's elements, Uranus had certainly been close to 34 Tauri in June and November 1690; and William Herschel himself thought that this identification might be valid.[20] Méchain and Baron Franz Xavier von Zach had meanwhile independently calculated Uranus's position with improved elements derived by Pierre Simon de La Place. They found, however, that the latitudes of that planet and star agreed but their longitudes differed by more than 2° - a discrepancy now regarded as being unacceptably large in comparison with the high degree of consistency found between calculations based on these same elements and the contemporary observations of Uranus's position. Consequently, Bode rejected 34 Tauri as a possible pre-discovery observation of Uranus in favour of "rather large perturbations, a change of the Sun's distance and of the node."[21]

There was, however, another possibility that required to be eliminated before this interpretation could be accepted; namely, that the discrepancy arose from errors in the reduction of Flamsteed's observation. This was investigated by Father Placidus Fixlmillner, who took account of aberration and nutation when reducing the heliocentric position of 34 Tauri to the time of Flamsteed's observation of 13 December [Old Style] 1690.[22] With this and the heliocentric position of Mayer 964, he was able to calculate orbital elements for Uranus that were compatible with Bradley's and Mayer's respective observations. Von Zach effected a further improvement by comparing the transit times of three bright stars and making a correction for the acceleration of Flamsteed's uncompensated pendulum clock.[23] Finally, following a suggestion from the Abbé Triesnecker in Vienna, Fixlmillner made a careful study of the errors in Flamsteed's mural quadrant and then corrected the transit times for these hitherto-ignored instrumental effects.[24] Calculations based on this last reduction and on Mayer's co-ordinates then appeared to indicate a slower mean motion for Uranus than that based upon contemporary observations.

Despite La Place's own acceptance of Mayer 964 as Uranus, a few practical astronomers still had reservations about the reliability of Mayer's observation, so it too had to be more care-fully studied before this apparent inconsistency in Uranus's mean motion could be accepted as real. Joseph Delambre, when preparing new tables of Uranus (1789), made a new reduction using data copied from Mayer's *Tagebuch* previously sent to Lalande by Lichtenberg. When he later came to discuss this in the second volume of his *Astronomie théorique et pratique* (Paris, 1814), he claimed that Mayer had incurred an error of 4 seconds in the timing of the transit and another of 3 seconds in aligning the plane of his mural quadrant with respect to the meridian. He also suggested, without reason or proof, that the observation in question had perhaps been made hastily and perhaps on a cloudy night. Now on the night of 25 September 1756, Mayer observed about a hundred stars in less than $3\frac{1}{2}$ hours at an average interval of only 40 seconds, and several of these were faint eighth and ninth magnitude stars. These facts in themselves testify to the excellence of the seeing conditions and surely prove that the time-interval of over 3 minutes that preceded the observation of Mayer 964 must have given him ample time to realign his telescope and to note the precise instant of transit. A strong counter-attack on the credibility of Delambre's con-clusions was made by von Zach,[25] who claimed that he was in posses-sion of Mayer's original manuscript catalogue.[26] He accused his French contemporary not merely of unjust criticism but also of incurring errors of transcription in transit times which had dis-torted his interpretation of Mayer's data. The most significant fact which Delambre did not even attempt to explain was that Mayer 964 remained missing from the place in the sky where it had been observed.

Towards the end of this same article, von Zach repeats a statement which he had already committed to print six years pre-viously in a historical review of astronomical advances during the first decade of the 19th century,[27] to the effect that Flamsteed and Pierre Charles Le Monnier had made pre-discovery observations of Uranus although the original observations had never come to light.[28] Flamsteed's observation of 34 Tauri on 23 December 1690 which Méchain, Fixlmillner and von Zach himself had previously investigated, had been taken not from Flamsteed's original observa-tion-book but from the posthumously-published star catalogue,[29] while Le Monnier's two observations of 20 and 23 January 1769 had already been reduced before being cited in Bode's *Jahrbuch*.[30] Yet Lalande, in introducing Delambre's tables of Uranus,[31] had remarked upon the high degree of precision with which these represented Flamsteed's, Mayer's, and Le Monnier's pre-discovery observations. Perhaps it was this stimulus from von Zach which encouraged Johann Karl Burckhardt[32] and Alexis Bouvard[33] to announce the results of their respective studies on Flamsteed's and Le Monnier's data soon afterwards.

In his lecture to the Institut de France on 16 December 1816, Burckhardt[34] attributes Flamsteed's second observation of Uranus on

22 March 1711/12 [Old Style] to the fact that he began observing a
few minutes later than he required to do in order to observe the
star ρ Leonis in which he was then interested. He also claims that
Flamsteed unknowingly observed the planet on three days - 21, 22,
27 February 1714/15 [Old Style] - when it happened to be in opposition to the Sun and in conjunction with Saturn, and yet again on 18
April 1715 [Old Style] when it transited several minutes later than
σ Leonis. He himself had not deliberately set out to find these
observations, since he had supposed that Bode would have detected
them in the course of his earlier researches.[35] Rather, he had
found them accidentally while comparing the star positions in some
southerly zones in La Caille's catalogue[36] and that of Caroline
Herschel[37] to those in Lalande's *Histoire céleste* (Paris, 1801).
His own elements of Uranus represented Flamsteed's first observation
of 1690 to about $\pm$ 1'.[38]

Bouvard explains that Le Monnier's frequent observations of
Uranus were due to his requiring a reliable catalogue of zodiacal
stars as a basis for the lunar observations that he wanted to make.
He had not identified any of these as the planet because he had not
needed to make day-to-day comparisons of their positions. Although
Le Monnier had made his observations from 1736 to 1780 with a mural
quadrant and poor-quality pendulum clock at the Observatory of the
Capuchines in Paris, his fifteen rough observation books were subsequently deposited at the Paris Observatory where Bouvard had
obtained access to them. Their quality was not good, since many
figures were scarcely decipherable, errors of several seconds often
occurred in the carelessly-recorded times of transit, and the
quadrant had not been placed exactly in the plane of the meridian.
Nevertheless, he managed to identify twelve instances (including
three previously noticed by Le Monnier himself) where Uranus had
been mistaken for a star, and listed the date, calculated mean time,
apparent right ascension, and apparent declination of each in a
table.[39] His careful reductions were founded upon Friedrich
Wilhelm Bessel's recently-published catalogue of James Bradley's
stars, in which Bessel himself detected a misidentified observation
of Uranus's declination made on 3 December 1753.[40] Two similar

unpublished observations by Bradley were to be discovered later,[41] but these were unreliable since they had been made before proper instrumentation was installed at the Greenwich Observatory.

In the introduction to his "Tables of Uranus" (1821), Bouvard[42] makes an explicit distinction between the seventeen other pre-discovery observations known at that time and the uninterrupted series made after Uranus's discovery forty years previously and concludes that, after due allowance had been made for the effects of planetary perturbations, it was still impossible to reconcile them. Because of this, he decided to adopt the elements for Uranus's orbit calculated from the later (more numerous and reliable) observations, leaving discordances of a few minutes of arc in several of the former for future astronomers to explain. Continued observations of this planet made during the following decade merely increased the complexity of the problem and, by 1834, led him to speculate on the existence of a disturbing body beyond the orbit of Uranus. This idea was not new. It had occurred as early as 1787 to Lexell, who reasoned that since certain comets return period- ically after many decades, gravity must extend to much greater distances than the orbit of Uranus.[43]   Bessel himself again makes explicit reference to it in a letter to Gauss dated 14 June 1824[44] and, after continued observations during the following decade served merely to emphasise that the discrepancies in the theory of Uranus's orbital motion could not be accounted for by errors in observation, he engaged a young student Friedrich Wilhelm Flemming to make a thorough study of the observed perturbations using this as the working hypothesis.[45] Owing to Flemming's nervous illness and Bessel's own preoccupation with other matters, this approach had not yielded any positive result before the young Cambridge undergraduate John Couch Adams independently resolved to undertake this laborious task in the summer of 1841.

The initial stimulus to Adams's investigation appears from his own testimony to have been the offer of a mathematical prize by the Göttingen Academy of Sciences, for the best analysis of the problem of Uranus's orbital motion;[46] and it is evident from another letter from Bessel to Gauss that it was he who instigated and Gauss

who arranged that such a prize should be awarded.[47]   Later, however,
in a lecture to the Royal Astronomical Society on 13 November 1846
at which the Astronomer Royal George Airy was present, Adams gives
him credit for having been the guiding spirit of his researches.[48]
An autobiographical history of Airy's involvement in this matter
was presented by the Astronomer Royal himself at the same meeting.[49]
He was first directly confronted with the problem in 1834 by the
Reverend T.J. Hussey, who had been talking to Bouvard during a visit
to Paris. On his return to England, Hussey wrote to Airy, telling
him that they were independently of the opinion - shared by others -
that there must be another planet superior to Uranus producing the
latter's unexplained perturbations, and asking him where one ought
to conduct a search for this disturbing body.

At that time, Airy had been highly sceptical of the idea, and
thought that it would be virtually impossible to discover such a
planet in any case. His considered opinion was that the unexplained
steady increase of Uranus's celestial longitude might simply result
from the adoption of too small a value for its mean distance from
the Sun. Thus he encouraged Alexis Bouvard's nephew (Eugène) to
explore this possibility. It was only after Eugène Bouvard's careful and determined effort to reconcile theory and observation
failed, that French and English astronomers finally became convinced
that the hypothesis of a superior planet beyond Uranus had to be
taken very seriously. Airy himself, however, was more concerned
with the question of whether a gravitational influence of this
nature was capable of explaining the error in Uranus's distance,
than with the existence of the disturbing body itself; and he
blamed Adams's failure to respond to his query on that issue for
his own tardiness in studying the manuscript containing Adams's
solution to the problem itself.[50]

Meanwhile, Urbain-Jean Joseph Le Verrier had begun to publish
a series of memoirs in the *Comptes rendus* dealing in depth with
various aspects of Uranus's motion, including the reduction of the
nineteen pre-discovery observations which Alexis Bouvard had
dismissed as uncertain. He too became firmly convinced in the
existence of another planet exterior to Uranus and, in his important

memoir of 1846,[51] makes a full analysis culminating in the prediction
of its longitude as about 325° on 1 January 1847. He communicated
this result to astronomers at the Berlin Observatory on 23 September
1846, who had no difficulty in discovering the new planet (Neptune)
that same evening.[52]

Now that the problem of Uranus's anomalous motion was satis-
factorily resolved, astronomers were able to re-compute its
positions for the times when the pre-discovery observations had been
made. In view of the care with which the reductions of Flamsteed's,
Mayer's and Bradley's observations had already been made, Le Verrier,
in his revision of Bouvard's tables, considered that it sufficed
merely to correct the mean times of Le Monnier's observations and
to revise the positions of the two made in 1750.[53] The results of
his corrections to Bouvard's reductions of these observations, his
own reductions of Flamsteed's observations, and Bessel's reduction
of Mayer's observation, are listed in Table I.[54] Bradley's three
observations have been omitted since only their right ascensions
can be determined with sufficient accuracy to be of value. In view
of the uncertainties inherent in these observed data, and the
difficulties involved in their reduction, little would be gained
from any further attempt to improve them.

Simon Newcomb was the first person to employ the general
theory of perturbations by Neptune (as well as by Jupiter and
Saturn) when determining the elements of Uranus's orbit.[55] He
calculated the values of the mean star positions from the "Star
Tables of the American Ephemeris" and the Greenwich Seven-Year
Catalogues for 1860 and 1867, and made the reductions to apparent
place with modern constants. His revised value for the mass of
Neptune was taken from an earlier publication by Safford.[56] The
results of a comparison between the pre-discovery observations of
Uranus and the positions obtained by numerical integration from
this more sophisticated theory, illustrated in Table II,[57] clearly
reveal that the level of agreement is exceptionally high. The fact
that a marginally poorer fit is obtained if Newcomb's own reductions
are substituted for those of Le Verrier[58] confirms both the validity
of the identifications and the rigour of the reductions. Taking

account of the three observations of Bradley, and a recent new
identification with a Flamsteed observation on 3 December [Old
Style] 1714,[59] we may safely conclude that Uranus was observed as a
star on no fewer than 22 occasions during the 81-year period between
December 1690 and December 1771, before William Herschel discovered
it and recognised it for what it was.

NOTES

1. *Astronomisches Jahrbuch für das Jahr 1784 nebst einer Sammlung
    der neuesten in die astronomischen Wissenschaften einschlag-
    enden Abhandlungen, Beobachtungen und Nachrichten. Mit
    Genehmhaltung der Königl. Akademie der Wissenschaften
    berechnet und herausgegeben von J.E. Bode, Astronom der
    Akademie* (Berlin, 1781).   The abbreviated title of this
    periodical, adopted below, is *Astr.Jahrb.*

2. "Ueber einen im gegenwärtigen 1781sten Jahre entdeckten
    beweglichen Stern, den man für einen jenseits der Saturns-
    bahn laufenden, und bisher noch unbekannt gebliebenen Planeten
    halten kann", *ibid.*, pp.210-20. In a footnote on p.211 of this
    article, Bode gives the following five variants of Herschel's
    surname that had appeared in the early French and English
    reports of the discovery: Mersthel, Hertschel, Herthel,
    Herrschell, and Hermstel.

3. *Ibid.*, p.218.

4. Brahe, T. *Historia coelestis jussu S.C.M. Ferd. III. edita
    complectens Observationes Astronomicas Varias ad historiam
    coelestem spectantes* (Augustae Vindelicorum, 1666).

5. Hevelius, J. *Machinae Coelestis pars prior* (Gedani, 1673);
    pars posterior (Gedani, 1679).

6. Flamsteed, J. *Historia Coelestis Britannicae*, 3 vols.
    (Londini, 1725).

7. Mayer, T. "Fixarum zodiacalium catalogus novus ex observatio-
    nibus Gottingensibus ad initium anni 1756 constructus", in
    Lichtenberg, G.C. (ed.), *Opera inedita Tobiae Mayeri I*
    (Gottingae, 1775), pp.49-74.

8. *Ibid.*, p.72

9. *Astr.Jahrb. für 1785* (Berlin, 1782), p.189.

10. The circumstances which resulted in this arrangement being
    made are described in the introduction to Forbes, E.G. (ed.)
    Tobias Mayer's Opera Inedita (London, 1971), p.12.

11. *Op.cit.*, note 7.

12. "Aus einen Schreiben des Herrn Prof. Lichtenberg an Herrn
    Hofrath Kästner, vom 1. Sept. 1781", *op.cit.*, note 9,
    p.192.

13. *Ibid*., p.190.

14. "Aus einem Schreiben des Herrn Herschel an mich" (London, 13
     August 1783), in Bode's *Astr.Jahrb. für 1786* (Berlin, 1783),
     p.258.

15. Méchain to Bode; Paris, 1 April 1784, in Bode's *Astr.Jahrb.
     für 1787* (Berlin, 1784), p.141.

16. Bode, J.E., "Fortgesetzte Bemerkungen über den neuen Planeten
     (Uranus)" in *op.cit*., note 14, pp.219-23.

17. *Op.cit*., note 18, p.246.

18. Bode, J.E., "Versuch eines Beweises, dass bereits Flamsteed
     im Jahr 1690 (so wie Tobias Mayer im Jahr 1756) den neuen
     Planeten beobachtet", *Astr.Jahrb. für 1787* (Berlin, 1784),
     pp.243-6.

19. Le Monnier, P.C., "Mémoire sur la Disparition de l'Étoile de
     la constellation du Taureau, que Flamstéed a placée dans son
     Catalogue, pour 1690, à 51^{d}46'50" de longitude, avec une
     latitude de 0^{d}5'$\frac{1}{3}$ méridionale", *Histoire de l'Académie Royale
     des Sciences, Année MDCCLXXXIV* (Paris, 1787), pp.353-4.

20. "Astronomische Beobachtungen und Nachrichten, von Herrn Prof.
     von Zach. Aus zweyen Briefen desselben an mich", Bode's *Astr.
     Jahrb. für 1788* (Berlin, 1785), pp.214-20. In a letter of
     21 May 1785, von Zach told Bode that the Graf von Brühl, de
     Luc, Aubert, and himself had all spent a night with Herschel
     at Datchet, discussing this and other matters; during which
     time, Herschel had expressed the opinion that 34 Tauri was
     "very probably the new planet" (*ibid*., p.214).

21. *Op.cit*., note 18, p.246.

22. Fixlmillner, P., "Untersuchung der Elemente der wahren
     Laufbahn des neuen Planeten", *ibid*.

23. *Op.cit*., note 20, pp.214-15.

24. Fixlmillner, P., "Ueber die Tafeln vom Uranus und neue
     Flemente der Bahn dieses Planeten", Bode's *Astr.Jahrb. für
     1792* (Berlin, 1789), pp.158-60.

25. Zach, F.X. von, "Ueber den von Tobias Mayer im Jahr 1756
     beobachteten Planeten Uranus", in B. von Lindenau and J.G.F.
     Bohnenberger (eds), *Zeitschrift für Astronomie und verwandte
     Wissenschaften, 3* (Tübingen, 1817), pp.3-22.

26. This would seem to be contradicted by Francis Baily in his
     introductory remarks on "Mayer's Catalogue of Stars", *Memoirs
     of the Royal Astronomical Society 4* (1831), 395-6.

27. Zach, F.X. von "Versuch einer geschichtlichen Darstellung der
     Fortschritte der Sternkunde im verlorenen Decennio", *Monat-
     liche Correspondenz zur Beförderung der Erd- und Himmels-
     Kunde 23* (1811), 205-56. See p.221.

28. *Op.cit*., note 25, p.20.

29.   *Op.cit.*, note 6; vol.2, p.86.

30.   *Astr.Jahrb. für 1793* (Berlin, 1790), p.20.

31.   Lalande, J.J. de, *Histoire Céleste Française, contenant les observations faites par plusieurs astronomes français 1* (Paris, 1801), p.188.

32.   Burckhardt, J.C., "Sur plusieurs observations de la planète Uranus qu'on trouve parmi les étoiles de Flamsteed", *Connaissance des Tems, ou des mouvemens célestes, à l'usage des astronomes et des navigateurs, pour l'an 1820* (1818), 408-9.

33.   Bouvard, A., "Extrait des registres des observations astronomiques faites par Lemonnier, à l'Observatoire des Capucins, rue Saint-Honoré, à Paris", *ibid. pour l'an 1821* (1819); *Additions*, 339-47.

34.   Burckhardt, J.C., "Sur l'opposition d'Uranus en 1715, et sur les résultats qu'on peut en tirer", *loc.cit.*, note 32, 410-12.

35.   *Ibid.*; footnote, p.410.

36.   La Caille, N.L. de, *Coelum Australe Stelliferum* (Paris, 1763). This catalogue contained a selection from a total of approximately 10,000 stars observed over a period of nineteen months at the Cape of Good Hope. The star-positions in it were later adjudged by Francis Baily (*Monthly Notices of the Royal Astronomical Society 5* (1833), 93) to be reliable only to within $\pm$ 30".

37.   Herschel, C. *Catalogue of Stars, taken from Mr. Flamsteed's observations ... and not inserted in the British Catalogue ... With ... remarks by William Herschel* (London, 1798).

38.   *Op.cit.*, note 34, p.412.

39.   *Op.cit.*, note 33, p.341.

40.   Bessel, F.W. *Fundamenta Astronomiae pro Anno MDCCIV deducta ex Observationibus viri incomparabilis James Bradley in Specula Astronomica Grenovicensi per Annos 1750-1762 institutis* (Regiomonti, 1818), p.283.   Mayer's observation of 25 September 1756 is reduced by Bessel on pp.284-5.

41.   Breen, H., "On early Observations of Uranus by Bradley", *Monthly Notices of the Royal Astronomical Society 24* (1864), 124-5.

42.   These are discussed by their author in the second part of the introduction to his *Tables Astronomiques publiées par le Bureau des Longitudes de France, contenant les Tables de Jupiter, de Saturne et d'Uranus, construites d'après la Théorie de la Méchanique Céleste* (Paris, 1821).

43.   Lexell, A.I. "Recherches sur la nouvelle Planète découverte par M. Herschel & nommée par lui Georgium Sidus", *Nova Acta Academiae Scientiarum Imperialis Petropolitanae, I* (1787), 78 and 79.

44.   *Briefwechsel zwischen Gauss und Bessel herausgegeben auf Veranlassung der Königlich Preussischen Akademie der Wissenschaften* (Leipzig, 1880), p.435.

45.   Bessel, F.W., "Ueber die Verbindung der astronomischen Beobachtungen mit der Astronomie", in H.C. Schumacher (ed.), *Populäre Vorlesungen über wissenschaftliche Gegenstände* (Hamburg, 1848), p.452.

46.   Adams, J.C., "An Explanation of the Observed Irregularities in the Motion of Uranus", *Memoirs of the Royal Astronomical Society 16* (1847), 429.

47.   Bessel to Gauss; Konigsberg, 8 November 1843, *op.cit.*, note 44 p.567.

48.   Adams, J.C. "On the Perturbations of Uranus", *The Nautical Almanac and Astronomical Ephemeris for the Year 1851, with an Appendix* (London, 1847), pp.265-93.

49.   *Monthly Notices of the Royal Astronomical Society 7* (1847), 121-52.

50.   *Ibid.*, p.131.

51.   Le Verrier, U.-J.J., "Recherches sur les Mouvements de la Planète Herschel, dite Uranus", *Connaissance des Temps ... pour l'An 1849* (Paris, 1846); *Additions*, pp.1-254.

52.   Detailed accounts of the circumstances surrounding this discovery are contained in A.F.O'D. Alexander, *The Planet Uranus* (London, 1965) and Edward M. Grosser, *The Discovery of Neptune* (Cambridge, Mass., 1962).

53.   *Op.cit.*, note 51, p.126.   A table in *ibid.*, p.129 contains the corrected mean times (based on the Paris meridian), the observed and calculated equatorial co-ordinates, and the residual differences in both these and the ecliptic co-ordinates.

54.   This is taken from Edgar W. Woolard, "Comparison of the Observations of Uranus previous to 1781 with theoretical positions obtained by numerical integration", *The Astronomical Journal 57* (1952), 35-38.   Compare his Table 1, p.36 with that in Le Verrier, *op.cit.*, note 51, p.129.

55.   Newcomb, S., "An Investigation of the Orbit of Uranus, with general Tables of its Motion", *Smithsonian Contributions to Knowledge* No. 262, vol.19 (Washington [D.C.], 1874).

56.   Safford, T.H., "On the Perturbations of Uranus and the Mass of Neptune", *Monthly Notices of the Royal Astronomical Society 22* (1862), 142-4.

57.   *Op.cit.*, note 54, p.37.

58.   This can be appreciated by an inspection of the results of Newcomb's reductions of Flamsteed's observations; namely, for cases 1, 3, 4, 5 in Table II: $-0\overset{s}{.}8$, $+5''$; $-0\overset{s}{.}1$, $-6''$; $+0\overset{s}{.}9$, $+2''$; and $-1\overset{s}{.}5$, $+10''$ respectively.

59. Rawlins, D., "A Long Lost Observation of Uranus: Flamsteed, 1714", *Publications of the Astronomical Society of the Pacific* *80* (1968), 217-19.

TABLE I.   OBSERVED POSITIONS OF URANUS

No.	Date (Gregorian)	Paris Mean Time	Observed App. α	App. δ	Observer
1	1690 Dec. 23	$9^h41^m25^s$	55°49'19".7	+19°35'14".4	Flamsteed
2	1712 Apr. 2	9 46 47	155 38 29.4	+11 00 55.2	Flamsteed
3	1715 Mar. 4	12 43 35	170 40 02.7	+ 4 54 27.9	Flamsteed
4	Mar. 10	12 19 02	170 25 39.3	+ 5 00 38.2	Flamsteed
5	Apr. 29	8 55 49	168 45 55.5	+ 5 41 53.1	Flamsteed
6	1750 Oct. 14	8 04 08	324 15 25.4	-15 01 41.3	Lemonnier
7	Dec. 3	4 48 51	324 34 52.4	-14 53 19.8	Lemonnier
8	1756 Sept.25	10 21 12	348 00 54.5	- 6 01 49.4	Mayer
9	1764 Jan. 15	5 12 00	12 37 39.0	+ 4 43 47.2	Lemonnier
10	1768 Dec. 27	7 38 42	31 26 52.0	+12 15 35.0	Lemonnier
11	Dec. 30	7 26 54	31 24 45.8	+12 14 55.4	Lemonnier
12	1769 Jan. 15	6 23 41	31 22 07.7	+12 14 26.0	Lemonnier
13	Jan. 16	6 19 46	31 22 23.4	+12 14 36.3	Lemonnier
14	Jan. 20	6 04 09	31 24 06.6	+12 15 19.0	Lemonnier
15	Jan. 21	6 00 16	31 24 33.8	+12 15 31.8	Lemonnier
16	Jan. 22	5 56 21	31 25 04.7	+12 15 45.7	Lemonnier
17	Jan. 23	5 52 26	31 25 28.5	+12 16 07.5	Lemonnier
18	1771 Dec. 18	9 06 53	43 58 06.0	+16 25 20.2	Lemonnier

TABLE II.   OBSERVED AND COMPUTED POSITIONS OF URANUS, AND RESIDUALS

No.	Date Reckoned from Greenwich Mean Noon	GMT Mean Noon	Observed α	δ	Computed α	δ	O-C Δα	Δδ
1	1690 Dec. 23	$9^h32^m04^s$	$3^h43^m15^s.87$	+19°35'02".7	$3^h43^m15^s.62$	+19°35'02".0	+.25	+0".7
2	1712 Apr. 2	9 37 26	10 22 31.92	+11 01 04.9	10 22 32.52	+11 01 03.2	-.60	+1.7
3	1715 Mar. 4	12 34 14	11 22 38.03	+ 4 54 42.6	11 22 38.07	+ 4 54 51.8	-.04	-9.2
4	Mar. 10	12 09 41	11 21 40.48	+ 5 00 52.8	11 21 40.36	+ 5 01 00.9	+.12	-8.1
5	Apr. 29	8 46 28	11 15 02.17	+ 5 42 03.9	11 15 01.58	+ 5 42 10.4	+.59	-6.5
6	1750 Oct. 14	7 54 47	21 36 59.98	-15 01 49.7	21 37 00.17	-15 01 45.7	-.19	-4.0
7	Dec. 3	4 39 30	21 38 18.91	-14 53 23.4	21 38 18.97	-14 53 26.7	-.06	+3.3
8	1756 Sept.25	10 11 51	23 12 03.04	- 6 01 54.9	23 12 03.11	- 6 01 51.9	-.07	-3.0
9	1764 Jan. 15	5 02 39	0 50 30.99	+ 4 43 47.7	0 50 31.23	+ 4 43 44.8	-.24	+2.9
10	1768 Dec. 27	7 29 21	2 05 45.74	+12 15 26.0	2 05 45.59	+12 15 25.0	+.15	+1.0
11	Dec. 30	7 17 33	2 05 37.38	+12 14 46.7	2 05 37.96	+12 14 50.4	-.58	-3.7
12	1769 Jan. 15	6 14 20	2 05 27.16	+12 14 19.0	2 05 27.34	+12 14 25.9	-.18	-6.9
13	Jan. 16	6 10 25	2 05 28.23	+12 14 29.4	2 05 28.39	+12 14 33.5	-.16	-4.1
14	Jan. 20	5 54 48	2 05 35.20	+12 15 12.5	2 05 34.60	+12 15 14.4	+.60	-1.9
15	Jan. 21	5 50 55	2 05 37.03	+12 15 25.4	2 05 36.65	+12 15 27.3	+.38	-1.9
16	Jan. 22	5 47 00	2 05 39.11	+12 15 39.5	2 05 38.90	+12 15 41.2	+.21	-1.7
17	Jan. 23	5 43 05	2 05 40.72	+12 16 01.3	2 05 41.35	+12 15 56.2	-.63	+5.1
18	1771 Dec. 18	8 57 32	2 55 50.57	+16 25 18.5	2 55 49.75	+16 25 16.5	+.82	+2.0

THE IMPACT ON ASTRONOMY OF THE DISCOVERY OF URANUS

ROBERT W. SMITH

MERSEYSIDE COUNTY MUSEUMS, LIVERPOOL, ENGLAND

The discovery of Uranus was a very much more important event
than the addition of one primary planet to the Solar System.
Indeed, it was to greatly influence future developments in a number
of diverse regions. In this paper we shall consider its impact in
three such areas: (1) the research on, and the acceptance of, the
Titius-Bode Law; (2) the search for, and subsequent discovery of,
a planet exterior to Uranus and (3) the direction of William
Herschel's own career.

If one reads any general astronomy textbook of today, there
is almost sure to be a reference to Bode's Law. Some of the more
historically minded authors even refer to it as the Titius-Bode
Law. In fact, although it was widely discussed only after the
discovery or Uranus, the Law was first suggested in 1766 by
Johann Daniel Titius, Professor of Mathematics at the University
of Wittenberg. In the course of his studies Titius had noticed in
a book by Christian Wolff a series of rounded off distances for
the planets. It struck Titius that if he added 1 and 5 units to
the distances for Mars and Saturn he could obtain a surprising
result since the distances could then be written in the following
form:

$$Mercury = 4 = 4$$
$$Venus = 4+3 = 7$$
$$Earth = 4+6 = 10$$
$$Mars = 4+12 = 16$$
$$? = 4+24 = 28$$
$$Jupiter = 4+48 = 52$$
$$Saturn = 4+96 = 100$$

But why should there be a gap between Mars and Jupiter? Titius

himself was moved to declare: "Why should the Lord Architect have
left the space empty? Not at all. Let us therefore assume that
this space without doubt belongs to the still undiscovered
satellites of Mars; let us also add that perhaps Jupiter still has
around itself some smaller ones which have not been sighted yet by
any telescope".[1] Six years later this progression was taken up by
the brilliant young German astronomer J. E. Bode. He argued that
the gap between Mars and Jupiter was filled not by an insignificant
moon (or moons), but by a planet. "Can you believe", Bode exclaimed
"that the Founder of the Universe had left this space empty?
Certainly not".[2]

     After Herschel's sighting of Uranus there had been some
delay before it had been generally admitted to be a planet. The
common assumption had been that Herschel had stumbled upon a comet.
Comets were central to the way astronomers in the 1780s perceived
the heavens.[3] They were seen as carriers of divine activity, some
natural philosophers speculated that they refuelled the Sun and
stars, and it was widely believed that they were capable of bearing
life.[4] Also the dramatic recovery of Halley's Comet in 1758 and 1759
had been hailed as a victory for Newtonian theory and had given a
further fillip to comet studies. Hence as comets were at the very
focus of astronomical thought and as no planet had been found in
recorded history, it had seemed almost unthinkable that a new planet
could be found.[5] This is certainly not to say that speculations on
the existence of further planets were absent before 1781 (witness
Bode himself). Nevertheless it took Herschel to break the shackles
restraining many, perhaps the overwhelming majority of, astronomers
to the tacit assumption that new planets were not to be observed.
In particular, the detection of Uranus persuaded some to look more
favourably on the Titius-Bode Law and its prediction of a planet
between Mars and Jupiter.

     Despite Bode's championing, the Law had stirred very little
interest before 1784. But in that year Bode had completed a
monograph on the recently discovered Uranus (the name Bode had
himself proposed for Herschel's planet).[6] Now Bode pointed out that

the distance of Uranus agreed very well with the progression and he
stressed that this could not be a chance result.

There was however a continuing problem with the Law: its
complete lack of any physical basis. This meant that some
astronomers remained unconvinced of its worth and that others, such
as Laplace, judged the Law to be no more than a peculiar game with
numbers.[7] Bode dismissed such objections. Indeed, he took every
opportunity to discuss the Law in the pages of the prestigious
Berliner Astronomisches Jahrbuch that he edited, and the Law was
convincing enough for Bode and a few colleagues to resolve upon a
systematic search for the elusive planet. Bode and von Zach, the
Law's two main advocates, as well as Schröter and Olbers, were among
the six astronomers who met at Schröter's home in September 1800 and
debated the best way of tracking it down. The favoured scheme relied
on the co-operation of 24 astronomers, each of whom was to
diligently scan 1/24th part of the sky along the zodiac.

Before the plan could be put into action it was overtaken by
events. Even before the invitation from Bode and his friends to
search a stretch of the zodiac had reached him, Giuseppe Piazzi, the
Director of the Palermo Observatory, had chanced upon on
1 January 1801 (in the course of constructing a new star catalogue)
the first of what would later be called minor planets.[8] To begin
with Piazzi did not think that he was following in Herschel's
footsteps, but he soon became convinced that the object that he had
found, named Ceres, was a new planet and the very planet that had
been predicted by the Law.[9] The mean distance of Ceres was in
excellent agreement with that forecast and so here was exceedingly
persuasive evidence the Law was far more than a mathematical
curiosity.

However, in March 1802 came the stunning news that Olbers had
found what seemed to be another planet, Pallas.[10] Moreover, Pallas was
soon calculated to be at almost the same mean distance as Ceres. Now
the Law's supporters faced the embarrassing task of explaining away
the fact that where there was supposed to be only one planet, there
were two. Was it possible to save the Law? Olbers was at first

inclined to think that Pallas was closer in nature to a comet than a planet. But then in June 1802 Olbers explained to Herschel his own daring hypothesis."What", he proposed, "if Ceres and Pallas were just a pair of fragments, or portions of a once greater planet which at one time occupied its proper place between Mars and Jupiter and was in size analogous to the outer planets, and perhaps millions of years ago, had, either through the impact of a comet, or from an internal explosion, burst into pieces". [11]

Herschel is usually thought of as spending most of his time observing nebulae and stars, but he was constantly interrupting his observing programs to observe the members of the Solar System. He was excited by Ceres and Pallas and himself provided observational evidence for Olber's hypothesis of a disintegrated or shattered planet. His micrometer measurements of Ceres had disclosed them to have unexpectedly small sizes: Ceres had a diameter of 162 miles and Pallas a diameter of 147 miles. [12] These findings had staggered Bode who was sure that Ceres was the eighth primary planet, and that Pallas was a special or exceptional planet, or perhaps comet, in its neighbourhood. By taking this position Bode was of course able to defend the Titius-Bode Law. Herschel was also impressed by the Law and he too sought to avoid its overturn. Herschel reasoned that if Ceres and Pallas were admitted to be primary planets then the Law would be wrecked, whereas if they were members of a different species, the Law's integrity could still be maintained. He judged that the comae that he glimpsed around them, the highly inclined orbits, and the sizes of Ceres and Pallas made it absurd to call them planets and so he coined the title 'asteroids' for them. [13]

The discoveries of Ceres and Pallas were significant not just because of their support for the Titius-Bode Law. They had in addition enlarged ideas on the construction of the Solar System, and in this they were continuing a process begun by the discovery of Uranus. Moreover, the glorious prospect of finding more asteroids motivated astronomers to observe the skies meticulously. Ceres and Pallas also brought home the message that future finds would probably be made as the result of careful comparisons of observations. Herschel had already demonstrated the power of methodical examinations of large areas of the heavens, but in 1802 he stressed that although he had made five reviews of the zodiac, both Ceres and Pallas had escaped him. He now pressed examiners of the zodiac to concentrate on the motions of the stars.[14] These sentiments were later echoed by the British Astronomer Royal, Nevil Maskelyne. In his opinion, "If astronomers would observe on two successive nights, they would run a chance of discovering new planets. Or if they observed stars twice in the same night, with an interval of 1, 2 or 3 hours, with a good equatorial instrument, they would find them out by their motion in the interval".[15]

Two more asteroids were soon detected, Juno in 1804 and Vesta in 1807. Vesta indeed being found as the result of a deliberate search for asteroids by Olbers. Both of the new asteroids were at similar distances to Ceres and Pallas.

The accurate predictions by the Titius-Bode Law of the mean distances of Uranus and the asteroids had established a very high reputation for it. The discovery of Ceres in 1801 had at the time even prompted some astronomers to speculate on the existence of a planet orbiting beyond Uranus at a distance in agreement with the Law. A tentative name, Ophion, was actually assigned to the as yet unseen eighth planet.[16] But three decades were to pass before there was good evidence for this surmise and we may look upon Uranus as a beacon brilliantly pointing the way to the new planet since evidence for its existence was to be provided by the motion of Uranus.

As Professor Forbes discusses elsewhere in this volume, searches through early records soon after the discovery of Uranus had turned up a number of observations of the planet, including one by Flamsteed as far back as 1690. Unfortunately the pre-1781 observations of Uranus did not mesh at all well with those made after the discovery, and no single elliptical orbit adequately represented the old and modern observations. One try to explain this anomaly was that a comet had struck Uranus close to the time

of its discovery, and that this collision had sensibly shifted its
orbit. But the mounting errors of the tables of Uranus calculated
solely from the post-discovery observations put paid to this
hypothesis.[17] Further, by 1832 G. B. Airy, the Director of Cambridge
Observatory and soon to be Astronomer Royal, was reporting that the
true position of Uranus on the sky differed by nearly half a minute
of arc from the then current tables of predicted positions. The very
size of this discrepancy meant that there was a growing pressure to
devise some sort of explanation of such seemingly bizarre behaviour.

By about 1840 the choice seemed to be between two
possibilities: firstly, that the law of gravitation might act in
some unexpected manner at the enormous distance of Uranus,
secondly, that a planet lay beyond Uranus and was causing the
disturbances. As to the first possibility, there was a tradition
of suspecting the correctness of, and even tinkering with, the
inverse square law. However, by the early 1840s Laplace had long
since shown to just about everybody's satisfaction that Newtonian
Theory could explain away any alleged irregularity in the orbits of
the planets and their satellites. Thus Newtonian Theory appeared to
nearly all astronomers to be the true system of the world and
beyond reproach, beyond reproach that is until all other means of
explaining the motion of Uranus had been thoroughly explored and
had failed.[18] In consequence, the generally favoured hypothesis was
that a planet was beyond Uranus and was perturbing Uranus. But if
an unseen planet was the cause, how could its location be
calculated?

Astronomers were familiar with the classical problem of
perturbations, but the problem they now had to tackle was the
entirely novel one of inverse perturbations in which one needed to
describe the disturbances of Uranus and then infer the mass and
orbital elements of the disturbing planet. Of course the story of
how John Couch Adams and U.J.J. Leverrier solved this forbidding
problem is now well known.[19] By the middle of 1846 Sir John Herschel,
who knew of the mathematical researches of both Adams and
Leverrier, looked upon the detection of a planet beyond Uranus as
merely a matter of time. He declared: "We see it as Columbus saw

America from the shores of Spain. Its movements have been felt
trembling along the far-reaching line of our analysis, with a
certainty hardly inferior to that of ocular demonstration".[20] In
England, James Challis at the Cambridge Observatory was trying to
provide just such an ocular demonstration. The observing method
that he was employing was that used in the searches for asteroids:
that is, scrutinising an area of sky on several occasions and
checking for sensible motions of any of the stars. But the method
was to prove to be too slow and Challis was beaten to the prize
of Neptune by the Berlin astronomers Galle and d'Arrest. Galle and
d'Arrest had the enormous advantage over Challis of having an
accurate star map of the zodiac complete to the ninth magnitude.
In 1830, when no asteroid had been found for some 23 years, Bessel
had suggested to the Berlin Academy of Sciences that such maps of
the zodiacal region be constructed. Galle and d'Arrest were thus
able simply to compare the area of sky around Leverrier's
predicted position for the planet with the stars on the
appropriate star map. Almost immediately they found Neptune.
Unfortunately for Challis, who had already been searching for two
months, this particular map had not even reached England on
23 September 1846, the date of Neptune's discovery.[21]

It is worth noting here that both Adams and Leverrier had
exploited the Titius-Bode Law in their calculations, and as
Uranus had done much to make this Law respectable,[22] it had pointed
the way to Neptune in more than one way. But it is ironical that
Neptune also led to the downfall of the Law, at least in its
simple form. Observations of the new planet soon disclosed a
distance much smaller than that predicted by the Law, roughly 30
astronomical units from the Sun instead of the predicted 38. This
was much too large for the Law in its simple form to remain
credible.

But so far I have not discussed what was the most far-
reaching implication of Herschel's sighting of Uranus. This was
that the discovery of the seventh planet marked a turning point in
Herschel's life. He became famous almost overnight and soon with

this fame came royal patronage.[23] Herschel was thereby freed from
the need to make his living as a musician, and he was able to
devote his entire energies to his passion for astronomy. Through
his programme of research on the natural history of the heavens he
broke with the traditional astronomy of the eighteenth century.[24] In
so doing, he was to alter fundamentally the concerns, goals and
techniques of astronomers. As a result, astronomy is still shaking
with the consequences of Herschel's discovery of Uranus.

Notes

1.   J.D. Titius von Wittenberg, translation into German from
      French of Betrachtung über die Natur, vom Herrn Karl Bonnett
      (Leipzig, 1766)7. The translation given is from p. 1014 of
      S. L. Jaki's "The early history of the Titius-Bode Law",
      American Journal of Physics, 40 (1972), 1014-1023. Jaki's
      paper gives a full account of the Law's early history. On
      this see also M. M. Nieto, The Titius-Bode Law of planetary
      distances: Its history and theory, (Oxford 1972)
      chapters 1-6.

2.   J.E. Bode, Anleitung zur Kenntniss des gestirnten Himmels
      (Hamburg, 1772) second edition, 462. See also Jaki, op.cit.
      ref.1, 1015.

3.   See S. Schaffer"'The Great Laboratories of the Universe':
      William Herschel on matter theory and planetary life"
      Journal for the History of Astronomy, 11 (1980), 81-111,
      pp. 96-100.

4.   See, for example, J. Ferguson, Astronomy explained upon
      Sir Isaac Newton's principles.... (London, 1756) 30.

5.   This point is made by S. Schaffer in "Uranus and the
      establishment of Herschel's astronomy", Journal for the
      History of Astronomy, 12 (1981), 11-26, p. 15.

6.   J. E. Bode, Von dem neu entdeckten Planeten, (Berlin, 1784).

7.   Jaki, op. cit. ref.1, 1019.

8.   See F. von Zach, "Über einen zwischen Mars und Jupiter
      längst vermutheten, nun wahrscheinlich entdecken neuen
      Hauptlaneten unseres Sonnen-Systems", Monatliche
      Correspondenz zur Berförderung der Erd-und Himmels-Kunde,
      3 (1801), 592-623.

9.   J. E. Bode Mémoires de l'Acadamie Royale des Sciences et
      Belles-Lettres.... (Berlin, 1804) 141.

10.   A helpful account of the discoveries of Ceres and Pallas is
      given in R. Grant, History of physical astronomy from the
      earliest ages to the middle of the nineteenth century,
      (London, 1852), 238-240.

11. Olbers to Herschel, 17 June 1802, quoted in C. Lubbock, The Herschel chronicle, (Cambridge, 1933) 272.

12. W. Herschel, "Observations on the two lately discovered celestial bodies", Philosophical Transactions, 92 (1802) 213-233.

13. Op. cit. ref. 12, 228.

14. Op. cit. ref. 12, 228-230.

15. Maskelyne to Gauss, 18 October 1804, quoted in E. Forbes "The correspondence between Carl Friedrich Gauss and the Rev. Nevil Maskelyne (1802-5)", Annals of science, 27 (1971) 213-237, p. 235.

16. L. Gilbert, "Ist der Ophion, (ein Planet der Uranusbahn,) ein noch unbekannter Weltkörper?", Annalen der Physik, 11 (1802), 482-485.

17. M. Grosser, The discovery of Neptune, (Cambridge, Mass., 1962) 47.

18. See J. Merleau-Ponty, "Laplace as Cosmologist" in Cosmology, History and Theology, (New York, 1977), edited by W. Yougrau and A. Breck, 283-291.

19. See Grosser, op. cit. ref. 17.

20. J. Herschel, letter on "LeVerrier's planet," Athenaeum, 3 October 1846, 109.

21. There had been earlier sightings but the planet had been mistaken for a star. It has recently been argued that Galileo saw Neptune: C. Kowal and S. Drake, "Galileo's observations of Neptune", Nature, 287 (1980) 311-313.

22. On the Law's high reputation see the comments of J.P. Nicholl Professor of Astronomy at the University of Glasgow: The phenomena and order of the Solar System, (Edinburgh, 1838) 238.

23. See Lubbock, op. cit. ref. 11, 78-132.

24. See the article by Dr. Hoskin in this volume as well as his William Herschel and the construction of the Heavens, (London, 1963) and Schaffer, op. cit. ref. 5.

PRESENT KNOWLEDGE OF URANUS

THE ORIGIN OF URANUS:
COMPOSITIONAL CONSIDERATIONS

M. Podolak

Dept. of Geophysics and Planetary Sciences

Tel-Aviv University

Ramat Aviv, Israel

ABSTRACT

Several cosmogonic theories are examined for their ability to explain the details of Uranus' composition as inferred from observations and interior models. Suggestions are made as to how future work may enable us to decide among competing scenarios.

INTRODUCTION

Ever since its discovery, by Herschel, in 1781, the planet Uranus has provided a useful testing ground for astronomical theories. In addition to providing evidence for the correctness of Newton's law of gravitation at distances greater than previously accessible from planetary observations, it also provided an excellent confirmation of the Titius-Bode "law". Even when the observations did not quite agree with theory, as when Uranus' orbit was found to deviate from the path predicted by Newtonian mechanics, the result was still a happy one, as Neptune was thereby discovered. While the respective status of Newton's and the Titius-Bode law are no longer considered pressing issues, there are other areas where Uranus may provide a useful test case.

In studies of the origin of the solar system, the processes encountered are generally complex and non-linear. Their full solution requires details of radiative transfer in a complex geometry, hydrodynamics, plasma processes and chemical kinetics. Clearly we are not yet in a position to model, in detail, the early history of the solar system. For this reason, theorists have had to rely

partly on their intuition to guide them through the nebulous regions
where detailed computations are not practical. While intuition is
a powerful aid, it is desirable to limit its use by relying on observations whenever possible, and constantly comparing the theory
with the object theorized about. For an object as complex as the
solar system, this is not easy, and so I will concentrate here on
just the planet Uranus, more particularly, its composition. In this
paper I would like to examine two broad categories of cosmogonic
theory to determine which of them provide a consistent picture for
Uranus' composition. The theories I will discuss are by no means
the only ones found in the literature, but they do have the following merits: ·

1.   they cover a wide range of physical processes,

2.   they are presently popular,

3.   they are conveniently summarized in review articles.

ACCRETIONAL THEORIES

Let us first consider the so-called accretional theories.
These theories assume that the sun was once surrounded by a not too
massive ($M \sim 0.1\ M_{sun}$) nebular disk. This disk may have been left
behind by a contracting protosun (Prentice, 1978) or acquired in
some other, unspecified way (Safronov, 1972). At any given position
in the disk, the temperature and pressure in the gas determine which
materials will be solid (Lewis, 1972; Grossman, 1972). These solids
accumulate into protoplanetary embryos. Near the sun only rocky
material will be solid, and terrestrial planets are formed. Further
away, the temperatures are lower, and ices begin to condense, most
notably H_2O, NH_3, and CH_4. With extra mass available, the embryo
grows to a larger size, and finally induces a dynamical instability
in the surrounding nebular gas. The gas falls onto the core
(embryo), and a gas giant planet results (Perri and Cameron, 1974;
Mizuno, 1980). There is, of course, the obvious difficulty of
explaining why Jupiter and Saturn accumulated so much more gas than

Uranus and Neptune, but that may, in fact, depend on details of the background pressure in the nebula (Mizuno, 1980). These details have not yet been adequately studied.

Based on this picture we would conclude that Uranus consists of a core of rock surrounded by ices, hydrogen, and helium. A number of models have been built based on this picture (Reynolds and Summers, 1965; Podolak and Cameron, 1974; Podolak, 1976; Hubbard and MacFarlane, 1980a; Podolak and Reynolds, 1981). In general these models assume that the nebular temperature was sufficiently low so that all of the available H_2O, NH_3 and CH_4 was frozen, and that these ices were accreted with the same efficiency as rock. One would then expect a solar ice to rock ratio, i.e. between 2.6 and 3.2 (Cameron, 1980; Zharkov and Trubitsyn, 1978). There are, however, two points regarding the details of the composition that require clarification.

If one examines the microwave spectrum of Uranus, one finds that at centimeter wavelengths, one sees brightness temperatures higher than one would expect if significant amounts of ammonia were present (Gulkis et al., 1978). The implication is that Uranus' atmosphere (and possibly the planet as a whole) is strongly depleted in NH_3. Prinn and Lewis (1973) proposed that the atmospheric depletion could be accounted for by the combination of NH_3 and H_2S to form NH_4SH. The NH_4SH would condense and form clouds, and since NH_4SH does not absorb at these wavelengths, the NH_3 would be effectively hidden. The problem with this proposal is that roughly equal amounts of sulfur and nitrogen are required, while the N/S ratio in solar composition is about 5 (Cameron, 1980). Recently Lewis and Prinn (1980) have suggested a mechanism for enhancing the abundance of sulfur relative to nitrogen above the solar value. They point out that at high temperature the equilibrium composition of solar mix gas will have N_2 as the most abundant nitrogen species, while most of the carbon will be in the form of CO, with about 1% in the form of CO_2. At lower temperatures the equilibrium species will be NH_3 and CH_4, but the kinetic pathways to these products are

extremely slow. Thus, in the vicinity of Uranus, most of the
nitrogen will still be in the form of N_2 if it once passed through
a high temperature region. It is thus possible to accrete solids
with a relatively low nitrogen abundance, and therefore a low N/S
ratio. In addition, since the vapor pressure of CO is approximately equal to that of N_2, it too will not condense in the vicinity of Uranus. Since substantial amounts of CO_2 ice will be
available, the C/N ratio will be higher than solar. In such a
situation, the additional nitrogen brought into the planet as N_2
during the hydrodynamic collapse phase is still not sufficient to
make the total N/S greater than one, and NH_4SH formation provides
an excellent mechanism for hiding the ammonia. In this scenario
oxygen will be accreted as H_2O, carbon mostly as CO_2, and nitrogen
as NH_4COONH_2 or NH_4NCO_3 [see Lewis and Prinn (1980) for details].
Most of the carbon (in the form of CO) will not be accreted as ice.
Indeed since much of the oxygen is tied up as CO, less is available
to form H_2O, and the expected ratio of ices to rock is about 0.5
(assuming $CO_2/CO \approx 0.01$).

It is useful to ask at this point what interior models can
say about the composition. In what follows I will base myself on
the results of Podolak and Reynolds (1981). In this work models
of the following type were considered. Uranus was assumed to consist of a core of rock surrounded by an envelope of H_2, He, H_2O,
NH_3, and CH_4. The H_2 and He were taken to be in the solar ratio,
and the ices (H_2O, NH_3, and CH_4) were in the solar ratio to each
other. The ratio of H_2 to ices in the envelope could be varied,
however. The resulting models are shown in Figure 1. Here I have
plotted the relationship between the oblateness and the quadrupole
moment of the gravitational field, J_2, for various Uranus models
(solid lines). The dashed lines show how the oblateness varies
with J_2 for various rotation periods according to the relation

$$\varepsilon = (1 + 1.5J_2)\ (1.5\ J_2 + m/2)$$

where m is the ratio between the centrifugal and gravitational accelerations:

$$m = \frac{\omega^2 R^3}{GM}$$

Here ω is the angular velocity of rotation, R is the equatorial radius of the Planet, M is its mass, and G is the Newtonian constant of gravitation. The rectangle in the figure encloses the range of measured values for ε and J_2 (Franklin _et al._, 1980; Nicholson _et al._, 1978). These data suggest a rotation period of 15-17 hrs., in agreement with a number of observers (Brown and Goody, 1980; Munch and Hippelein, 1980; see also Goody this volume). There is, however, another value found in the literature, and that is one of 24 hrs. (Smith and Slavsky, 1979; Belton _et al._, 1980). This is indicated in the figure by the error bars on the 24 hr line.

As for the models themselves, U1 has a core of rock, and an envelope of solar composition. As one goes towards U6, the models

Figure 1. Oblateness and J_2 for various Uranus models. Observational limits on these quantities are also shown.

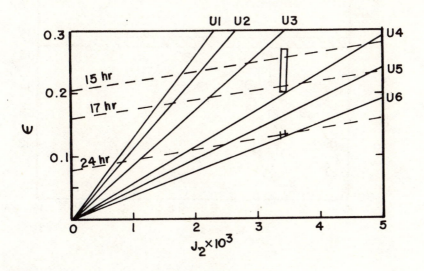

contain progressively more ices in the envelope, so that for U1 the
ice to rock ratio is about 0.015, while for U6 the ratio is about
4.2.   The relative amounts of $H_2 + H_e$, ice, and rock for the different models are shown in figure 2.   The shaded area in the figure
shows the range of estimates for the solar value of the ice to rock
ratio.   Comparison of figures 1 and 2 shows that if we accept the
24 hr period, this implies a model like U6, and hence a solar ice
to rock ratio.   This, in turn, implies that the material in the
vicinity of Uranus had been in chemical equilibrium at the time of
accretion.   On the other hand, from consideration of the kinetics
we would expect an ice to rock ratio of about 0.5, i.e. a model
between U3 and U4.   Such a model requires a rotation period of about
16 hrs. to fit the observations, and thus passes right through the
rectangle in figure 1.   It is worthwhile to note that while a 24 hr
period requires a solar ice to rock ratio, a 16 hr. period does not

Figure 2.   Amounts of rock, ice, and H_2 + He
for various Uranus models.

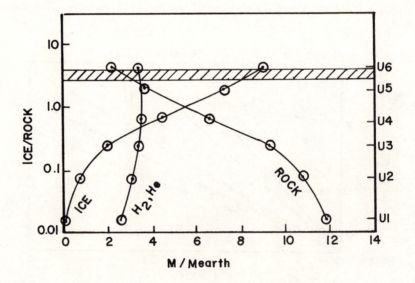

necessarily imply that the ratio is less. One can, for example, construct a planet with a solar ice to rock ratio, placing some of the ice in a shell around the core, and mixing the rest through the envelope (see the primed models in Podolak and Reynolds, 1981). In such a case the moment of inertia will be sufficiently low so that the observed J_2 can be fit with a 16 hr period. Thus a period in the vicinity of 16 hrs., while consistent with a non-equilibrium solar nebula, is not sufficient to rule out a solar ice to rock ratio. It is possible that a good value for J_4, the next term in the expansion of the gravitational field, will provide useful con-straints once the rotation period has been unambiguously determined.

Finally there is one other compositional issue that has been brought to light by the work of Hubbard and MacFarlane (1980b). This regards the deuterium to hydrogen ratio. For a gas of given (solar, say) composition, the relative amount of deuterium that will be confined in ices rather than in the hydrogen gas depends on temperature (Richet _et al_., 1977). Thus at low temperatures (<200K) the ices will contain between 10 and 10^3 times more deuterium than at high temperatures. Thus Uranus (and Neptune), which accreted a large fraction of ices relative to H_2 should have a D/H ratio con-siderably enhanced over the solar value of ~ 20 ppm. (Cameron, 1980; Black, 1973). In fact, the observations indicate a D/H ratio some 1–3 times the solar value (Macy and Smith, 1978; Trafton and Ramsay, 1980). Hubbard and MacFarlane evaluate the D/H ratio expected for Uranus models with the full complement of ices relative to rock, and find that the ratio should be about six times the solar value if the original ices were enhanced in deuterium by only a factor of 10. If the solar nebula were in equilibrium, then at the tempera-tures prevailing in the vicinity of Uranus enhancements of several hundred should be found. That the observed value is much lower in-dicates that either equilibrium is not achieved, or that the lower and upper parts of the atmosphere are not mixed. It is hard to

imagine that throughout its history Uranus never passed through a
high temperature convective stage. In addition, the release of
radioactive heat from the rock core (assuming chondritic composition) should be enough to drive mild convection even today
(Danielson, 1974; private communication). A non-equilibrium process is thus implied.

Two processes suggest themselves. First, due to the slow
kinetics at low temperatures, the material does not completely
equilibrate, and the ices in Uranus' vicinity did not acquire their
full complement of deuterium. A second non-equilibrium process is
the one mentioned earlier, i.e. that nitrogen does not enter the
planet as NH_3 ice, and carbon as CH_4, but rather as CO_2 ice. This
has two effects: 1. a smaller fraction of the total planet mass is
in the form of ice (the H_2/ice ratio is higher);2. the ice has a
smaller mass fraction of hydrogen, and therefore of deuterium.
These last two considerations alone are sufficient to reduce the
expected D/H ratio to about three times the solar value (not inconsistent with the observed value). We see therefore that an accretional theory of Uranus' formation can account for:

1. The enhancement of rock and ices relative to solar
    composition.
2. The apparent absence of nitrogen.
3. The D/H ratio.

GIANT PROTOPLANET THEORIES

A second group of theories results in the formation of giant
gaseous protoplanets. These come about in various ways. In the
encounter theory of Woolfson (1978a,b) the sun passes close to an
extended protostar, and draws off a filament of stellar material.
Since the filament comes from the protostar, the angular momentum
constraint of Russel (1935) doesn't arise, and since the material
is at much lower temperatures than solar material, the objection

of Spitzer (1939) doesn't apply. This theory is of special interest since, if it is correct, it implies that planetary systems are much less common than is generally assumed. According to the encounter picture, then, the filament, which is unstable, breaks up into several smaller masses. These protoplanets then evolve into the planetary system we see today. The important point for us is the fact that the planets originate as giant gaseous protoplanets whose overall composition is solar.

A second scenario which also results in the formation of giant protoplanets is the floccule theory of McCrea (1978). Here one treats a turbulent cloud of approximately one solar mass. The turbulence is supersonic, and the resulting eddies are approximated by so called "floccules" which act (and interact) like spheres orbiting about a common center of mass. Initially the inclinations of the orbit are random, as are the eccentricities and the sense of the orbit (prograde or retrograde). If two floccules collide with nearly zero total angular momentum, they fall towards the center of the cloud, and the sun gradually develops. If the net angular momentum is not zero, this new, larger floccule continues to orbit. When the mass of one of the floccules reaches the Jeans mass, it begins to contract, and from this point the evolution proceeds as it would for a giant gaseous protoplanet. The floccule theory is especially interesting since it combines aspects of giant planet theories with those of accretional theories.

A third scenario which results in the formation of giant gaseous protoplanets is that proposed by Cameron (1978a,b). He considers the evolution of a massive (twice the solar mass) solar nebula. As the collapse of the solar nebula proceeds, a central condensation, the protosun, is formed, surrounded by a gaseous disk. As material continues to fall onto the disk it becomes unstable, and a ring is formed. In some sense this ring is similar to the filament of the encounter theory. It too breaks up into a number of subcondensations which interact with each other, generally either colliding and coalescing, or escaping from the system. The remaining sub-condensation continues to evolve as a giant proto-planet. As the gas continues to fall onto the disk, additional rings and additional protoplanets are formed. Like the floccule theory this theory implies that stars and planets are formed in the same process and that planetary systems should be quite common. Unlike the floccule theory, it treats a number of different processes and, on the whole, presents a rather different picture.

In all three cases we start with protoplanets of solar composition. In order to form Uranus there must be some sort of loss mechanism for hydrogen and helium. One such mechanism has been proposed by Handbury and Williams (1975). They point out that for the temperatures and gravitational potentials found at the surface of giant protoplanets, cooling through Jeans escape is a much more efficient process than cooling through radiation into space. They suggest that if all of the gravitational energy released by settling of grains towards the planetary core were removed by Jeans escape of hydrogen and helium, one could explain the present densities of Uranus and Neptune. Indeed this mechanism is just the one suggested by Woolfson (1978b) as a possible way for the proto-Uranus and proto-Neptune of his theory to lose their excess hydrogen and helium. The difficulty with such a mechanism is the nitrogen abundance. If it is in the form of NH_3, it does not escape, and the N/H ratio should be much higher than solar. Similarly, if it was in the form of N_2, it would have too high a molecular weight to escape efficiently via the Jeans process. Again one would expect a N/H ratio much higher than solar in contradiction with the observations (Gulkis _et al._, 1978). Here NH_4SH formation would not resolve the problem, because in solar composition the S/N ratio is too small. It is possible that some other mechanism exists for removing the NH_3 from the atmosphere, but no concrete proposal has yet been presented.

In the theory of McCrea a rotational instability results in the shedding of material as the protoplanet contracts. If the grains settle through the gas quickly, only gases are shed, and the resulting rocky core is the precursor of a terrestrial planet. If the grains settle slowly through the gas, both grains and gas are shed, and the resulting planet (with its nearly solar composition) is the precursor of one of the Jovian planets. McCrea suggests that something intermediate between these two extremes would describe the formation of Uranus and Neptune. This would account for the low (relative to rock and ice) hydrogen and helium abundance. The expected values of the N/H and S/N ratios will again be dependent on the form of nitrogen in the protoplanet. If it is in the form of NH_3 it will settle with the grains and the N/H ratio will be greater than solar while the S/N ratio will be solar, thus there will again not be enough sulfur for complete NH_4SH formation. If the nitrogen is in the form of N_2, the N/H ratio will be solar and the S/N ratio will be considerably higher than solar, so that enough sulfur will be available for complete conversion of NH_3 to NH_4SH. Clearly detailed modelling of giant protoplanet evolution for this theory is an important next step. In addition it would be important to show why the rate of settling of the grains should differ in different parts of the solar system.

Cameron's theory considers two major stages for protoplanet evolution. The first occurs when the run of pressures and temperatures inside the protoplanet passes through the liquid part of the phase diagram of some rocky or metallic component (iron for example). When this occurs, droplets are formed which grow rapidly (Slattery, 1978) and fall towards the center to form a core. The second stage involves tidal stripping of the envelope by the sun. The protoplanetary core, a terrestrial planet, remains. If the protoplanet is too far from the sun for tidal stripping to occur, the envelope contracts until temperatures at the core boundary become high enough

for hydrogen to dissociate. The envelope then becomes unstable and
a hydrodynamic collapse ensues. A giant planet is thus formed.
The difficulty is that even for Jupiter and Saturn, and all the more
so for Uranus and Neptune, there is an enhancement of rock and ice
above the solar value relative to hydrogen. It is important to re-
call that in this theory the protoplanet is imbedded in the solar
nebula, and the protoplanetary envelope joins to the nebula itself.
It may be that transport of material between the envelope and the
nebula occurs right up to the time of envelope collapse (see, however, Cameron, 1979). If such mixing occurs, then the amount of gas
participating in the final collapse may vary depending on the background conditions in the nebula. As Cameron (1978a,b) has pointed
out, there may be a nebular "breeze" which will greatly reduce the
mass of the nebula in the last stages of evolution. If the Uranus
and Neptune protoplanets evolved more slowly than their Jovian and
Kronian counterparts, they would have found themselves in a nebula
with very different physical conditions at the time of collapse.
This could easily affect the size of the collapsing envelope.

Finally, the idea of mixing between the protoplanet and the
nebula bears on the problem of the N/H ratio. Here again we have
to turn to protoplanet models to determine the form that nitrogen
will be in. The most detailed models to date of the giant protoplanets are those by DeCampli and Cameron (1979). They find that
interior temperatures of more than 10^3K are reached as the protoplanet evolves. For the relevant pressures the equilibrium form of
nitrogen is N_2. In addition, the convection they find is sufficiently vigorous that in the lower temperature regions where the
kinetics is still sufficiently slow (Norris, 1980), the major
nitrogen species is still N_2. In this case the following scenario
suggests itself. Suppose that gases move between the protoplanet
and the nebula, but that near the top of the protoplanetary atmosphere there is a cold trap. H_2 and He will pass through, and if
the temperature is sufficiently high N_2 will pass through as well.

H_2O (or H_2S for that matter) will not, however. When the final collapse of the atmosphere occurs, there may well be a high H_2O/H_2
ratio, and an S/N ratio larger than one. An ice rich envelope will
therefore be formed where NH_4SH formation will hide any NH_3 produced
when the planet reequilibrates after collapse.

The question that remains concerns the status of carbon. If
most of the carbon is in the form of CO, it should pass through the
cold trap if N_2 does, since their vapor pressures are similar. The
resulting planetary composition(ice/rock $\approx$ 0.5) will be similar to
that expected for the non-equilibrium accretion model. If the carbon is in the form of CH_4 (as happens for the ranges of temperature
and pressure found in some of the models of DeCampli and Cameron),
then it has a smaller chance of passing through the cold trap,
since its vapor pressure is lower. If it does pass through, then
the expected ice/rock ratio will be about 2, since more oxygen is
now available to form H_2O. If the CH_4 does not pass through, the
expected ice/rock ratio is about 3. As pointed out earlier, models
of Uranus and Neptune do not as yet provide definitive values for
this quantity. It should be possible, by more careful studies of
the evolution of imbedded giant gaseous protoplanets, and more detailed modeling of the present day interiors of Uranus and Neptune,
combined with an improved knowledge of their gravitational field
to decide if a massive solar nebula is indeed a viable scenario for
planet formation. Finally it is important to point out that all the
giant planet scenarios start with a solar D/H ratio, which should
be maintained throughout their evolution. Within the uncertainties
in the measurements, this agrees with the observed value.

CONCLUSIONS

In this paper I have tried to show how the composition of Uranus can be used as a benchmark to test theories of the origin of the solar system. Since the details of Uranus' composition are still not unambiguously determined, it is impossible to draw firm conclusions, nontheless it does seem that accretional theories re-produce the low hydrogen abundance together with a low N/H and N/S ratio in a more straightforward way. Thus they can explain Uranus' high density (relative to Jupiter and Saturn) and the apparent absence of NH_3. Giant protoplanet theories are not ob-viously inconsistent with these data, but neither do they show clearly how such a situation developed. With regard to the D/H ratio, giant protoplanet theories predict a ratio near the solar value, while accretional theories have difficulty in keeping the ratio even as low as three times the solar value. Overall, it seems that accretional theories provide a somewhat more comfortable frame-work for understanding the composition of Uranus, but it is clear that it will require considerable work before we can claim we truly understand the planet's origin.

ACKNOWLEDGEMENTS

I am indebted to R.T. Reynolds, A.G.W. Cameron, and Yu.Mekler for many interesting discussions. In particular I want to thank the students in my various classes who by their astute questions helped to guide my thinking. This work was supported, in part, by NASA Grant NCA 2-OR-340-002.

REFERENCES

Belton, M.J.S., Wallace, L., Hayes, S.H., and Price, M.J. (1980).
Neptune's rotation period: A correction and a speculation on the
differences between photometric and spectroscopic results.
Icarus 42, 71-78.

Black, D.C. (1973). Deuterium in the early solar system. *Icarus*
19, 154-159.

Brown, R.A., and Goody, R.M. (1980). The rotation of Uranus II.
Astrophys. J. 235, 1066-1070.

Cameron, A.G.W. (1978a). The primitive solar accretion disk and the
formation of the planets. In *The Origin of the Solar System*
(S.F. Dermott, ed) pp. 49-74. John Wiley & Sons, New York.

Cameron, A.G.W. (1978b). Physics of the primitive solar nebula and
of giant gaseous protoplanets. In *Protostars and Protoplanets*
(T. Gehrels, ed.) pp. 453-487. Univ. of Arizona Press, Tucson.

Cameron, A.G.W. (1979). The interaction between giant gaseous protoplanets and the primitive solar nebula. *Moon and Planets* 21,
173-183.

Cameron, A.G.W. (1980). Elementary and nuclidic abundances in the
solar system. To appear in *A Festschrift in Honor of Willy
Fowler's 70th Birthday*.

DeCampli, W., and Cameron, A.G.W. (1979). Structure and evolution
of isolated giant gaseous protoplanets. *Icarus* 38, 367-391.

Elliot, J.L., French, R.G., Frogel, J., Elias, J.H., Mink, D.J.,
and Liller, W. (1981). Orbits of nine Uranian rings. *Astron. J.*
in press.

Franklin, F.A., Avis, C.C., Columbo, C., and Shapiro, I.I. (1980).
Geometric oblateness of Uranus. *Astrophys. J.* 236, 1031-1034.

Goody, R.M. (1981). The rotation of Uranus. This volume.

Grossman, L. (1972). Condensation in the primitive solar nebula.
Geochim. et Cosmochim. Acta 36, 597-619.

Gulkis, S., Janssen, M.A., and Olsen, E.T. (1978). Evidence for the
depletion of ammonia in the Uranus atmosphere. *Icarus* 34, 10-19.

Handbury, M.J., and Williams, I.P. (1975). The formation of the
outer planets. *Astrophys. and Space Sci.* 38, 29-37.

Hubbard, W.B., and MacFarlane, J.J. (1980a). Structure and evolution of Uranus and Neptune. *J. Geophys. Res.* 85, 225-234.

Hubbard, W.B., and MacFarlane, J.J. (1980b). Theoretical predictions of deuterium abundances in the Jovian planets. *Icarus*, in
press.

Lewis, J.S. (1972). Low temperature condensation from the solar
nebula. *Icarus* 16, 241-252.

Lewis, J.S., and Prinn, R.G. (1980). Kinetic inhibition of CO and
N_2 reduction in the solar nebula. *Astrophys. J.* 238, 357-364.

McCrea, W.H. (1978). The formation of the solar system: A protoplanet theory. In *The Origin of the Solar System* (S.F. Dermott,
  ed.) pp. 75–110, John Wiley & Sons, New York.
Macy, W. Jr., and Smith, W.H. (1978). Detection of HD on Saturn and
  Uranus and the D/H ratio. *Astrophys. J.* 222, L73–L75.
Mizuno, H. (1980). Formation of the giant planets. *Prog. Theore-
  tical Phys.* 64, 544–557.
Munch, G., and Hippelein, H. (1980). The effects of seeing on the
  reflected spectrum of Uranus and Neptune. *Astron. Astrophys.* 81,
  189–197.
Nicholson, P.D., Persson, S.E., Matthews, K., Goldreich, P., and
  Neugebauer, G. (1978). The rings of Uranus: Results of the 1978
  April 10 occultation. *Astron. J.* 83, 1240–1248.
Norris, T.L., (1980). Kinetic model of ammonia synthesis in the
  solar nebula. *Earth and Planet. Sci. Lett.* 47, 43–50.
Perri, F., and Cameron, A.G.W. (1974). Hydrodynamic instability of
  the solar nebula in the presence of a planetary core. *Icarus* 22,
  416–425.
Podolak, M. (1976). Methane rich models of Uranus. *Icarus* 27, 473–
  478.
Podolak, M., and Cameron, A.G.W. (1974). Models of the giant
  planets. *Icarus* 22, 123–148.
Podolak, M., and Reynolds, R.T. (1981). On the structure of Uranus
  and Neptune. *Icarus* 45, in press.
Prentice, A.J.R. (1978). Towards a modern Laplacian theory for the
  formation of the solar system. In *The Origin of the Solar System*
  (S.F. Dermott, ed.) pp. 111–162. John Wiley & Sons, New York.
Prinn, R.G., and Lewis, J.S. (1973). Uranus atmosphere structure
  and composition. *Astrophys. J.* 179, 333–342.
Reynolds, R.T., and Summers, A.L. (1965). Models of Uranus and
  Neptune. *J. Geophys. Res.* 70, 199–208.
 ichet, P., Bottinga, Y., and Javoy, M.A. (1977). A review of hydro-
  gen, carbon, nitrogen, oxygen, sulfur, and chlorine stable isotope
  fractionation among gaseous molecules. *Rev. E. & P. Sci.* 5, 65–
  110.
Russel, N.N. (1935). *The Solar System and its Origin.* Macmillan,
  New York.
Safronov, V.S. (1972). *Evolution of the Protoplanetary Cloud and
  Formation of the Earth and Planets,* Nauka, Moscow. Transl. from
  Russian Israel Program for Scientific Translation, Jerusalem.
Slattery, W. (1978). Protoplanetary core formation by rainout of
  iron drops. *Moon and Planets* 19, 443–456.
Smith, H.J., and Slavsky, D.B. (1979). Rotation period of Uranus.
  Bull. Amer. Astron. Soc. 11, 568.

Spitzer, L., Jr. (1939). The dissipation of planetary filaments.
  Astrophys. J. 90, 675–688.
Trafton, L., and Ramsay, D.A. (1980). The D/H ratio in the at-
  mosphere of Uranus: Detection of the $R_5(1)$ line of HD. *Icarus,*
  41, 423–429.
Woolfson, M.M. (1978a). The capture theory and the origin of the
  solar system. In *The Origin of the Solar System* (S.F. Dermott,
  ed.) pp. 179–198. John Wiley & Sons, New York.
Woolfson, M.M. (1978b). The evolution of the solar system. In
  The Origin of the Solar System (S.F. Dermott, ed.) pp. 199–217.
Zharkov, V.N., and Trubitsyn, V.P. (1978). *Physics of Planetary
  Interiors* (W.B. Hubbard, Transl. and ed.) Pachart, Tucson, Ariz.

Internal Structure of Uranus

J. J. MacFarlane and W. B. Hubbard

Department of Planetary Sciences

Lunar and Planetary Laboratory

University of Arizona

Tucson, Arizona 85721

ABSTRACT

We present an updated study of Uranus interior models using current information about the planet's gravity field and rotation rate. The most plausible model, both from the point of view of recent data and cosmogony, has a central core of iron and magnesium silicates, an outer envelope of liquid water, methane, and ammonia, and a deep "atmosphere" of almost four earth masses of hydrogen, helium, and methane. The "atmosphere" contains a gravitationally nonnegligible amount of methane -- about 40% by mass. All plausible models are most consistent with a rotation period of $\sim$15 to 16 hours.

I. INTRODUCTION

The goal of studies of the interior of Uranus is to achieve a synthesis of data on the planet's gravitational and magnetic fields, average density, atmospheric composition, heat flow, and various cosmogonical considerations. Recent years have seen a reduction in the number of possible interior models, although many uncertainties still remain. The purpose of this paper is to summarize a number of recent developments and to indicate the status of Uranus interior models in the context of a number of observational constraints, as of early 1981. A set of Uranus and Neptune models were presented by Hubbard and MacFarlane (1980; HM hereafter); this paper serves as an update on several important results since that time. Due to space limitations, this paper cannot also serve as a review paper and so we will be unable to discuss in any detail such important studies of Uranus' interior as Reynolds and Summers (1965), Podolak and Cameron

(1974), Podolak (1976), and Podolak and Reynolds (1981).

Although preferably one should first derive an interior model and then draw cosmogonical conclusions from it, the Uranus problem is still so ill-constrained that it is preferable to limit interior models to those consistent with a plausible cosmogony. Thus we consider a scenario for formation of Uranus (and Neptune) roughly as follows. Cooling of an initially hot solar-composition gas to temperatures below $\sim 1400^{\circ}$K will result in the condensation of iron and magnesium silicates ("rock"). At still lower temperatures, say $\lesssim 150^{\circ}$K, various abundant species such as H_2O and then NH_3 and CH_4 will likewise form solid condensates ("ice"), leaving behind a gas phase composed principally of H_2 and He in solar proportions. Now it is unclear whether material in Uranus' formation zone was ever heated to temperatures high enough to vaporize silicates, although H_2 and He will of course never condense and thus would be expected to be fractionated with respect to condensibles. It is assumed (Mizuno 1980) that the condensed species in Uranus' formation zone will aggregate to form planetesimals and then a planetary core with a mass of at least several M_E (one earth mass = M_E). We would expect this core to be made up of "rock" and "ice", with the ratio of "ice" to "rock" (I/R) depending on the precise chemical state of the condensibles and on whether CH_4, NH_3, etc. have completely condensed. For solar composition and complete condensation of "ice" and "rock", $I/R \simeq 3$ (HM).

According to Mizuno's (1980) calculations, a small amount proportionally of H_2-He, $\sim$ few M_E, will then be captured by the protoplanetary core in Uranus' formation zone. Cosmogonically then, we expect that the hydrogen-rich atmosphere of Uranus is by mass only a small fraction of the planet, although this result may have benefited by hindsight from earlier interior models such as Reynolds and Summers (1965).

Since "ice" will condense after "rock", it seems most plausible that $I/R \leq 3$ in Uranus' interior. Models with proportionally more "ice" than this limiting value would need to be produced by an initially chemically inhomogeneous nebula. In any case, we would normally expect "ice" and "rock" to be separated in the interior of Uranus.   For an adiabatic temperature distribution in the planet, which seems most plausible because of the low thermal conductivity of the planetary material (Zharkov and Trubitsyn, 1972), we have typical interior temperatures $\sim$5000K at pressures of several megabars (HM).   According to estimates by Hubbard (1981), iron and magnesium silicates will tend to be solid under these conditions but "ice" will be liquid.   In view of the great density contrast between "ice" and "rock", it seems inevitable that a "rock" core will be formed.

The above considerations led HM to consider three-layer models for Uranus and Neptune, consisting of a central "rock" core, a mantle composed of "ice", and a deep atmosphere composed primarily of hydrogen and helium. These models were the most centrally condensed ones which had been proposed, having dimensionless moment of inertia factors $C/Ma^2 \simeq 0.20$ (C = polar moment of inertia, M = planetary mass, a = equatorial radius at 1 bar pressure). In contrast, Jupiter has $C/Ma^2 \simeq 0.26$ and for Saturn $C/Ma^2 \simeq 0.23$.   We must stress that there is no "similarity law" for the structure of Jovian planets. In Jupiter and Saturn, C/Ma^2 is basically determined by the structure of the deep hydrogen-rich envelope and by a relatively small core.   The detailed structure of the core plays no role.   For Uranus and Neptune, however, the structure of the core is as important as the structure of the hydrogen-rich envelope, and it makes a significant difference whether or not the "ice" and "rock" components are separated.

In the following, we will consider some new observational constraints and some resulting variations on the HM models.

II.   CONSTRAINTS ON MODELS

The principle constraints on interior models are M, a, and the dimensionless zonal harmonics of the gravity field J_{2n}, defined by

$$V(r,\theta) = \frac{GM}{r} [1 - \sum_{n=1}^{\infty} J_{2n}(a/r)^{2n} P_{2n}(\cos\theta)], \tag{1}$$

where V is the planet's external gravitational potential as a function of distance r from the center of mass and angle θ from the rotation axis, G is the gravitational constant, and P_{2n} are Legendre polynomials. The best available value of a is 25,900 $\pm$ 300 km (Danielson, et al. 1972) obtained from an analysis of Stratoscope images. Occultation measurements give a' = 26,200 $\pm$ 100 km (Elliot, et al. 1981) for the equatorial radius at the occultation level which is much higher than the 1-bar level. In fact the above values of a and a' are consistent. The error bar in a' has been increased above the formal value of 30 km to allow for possible systematic errors. Although it is customary to define the J_{2n} by using the 1-bar equatorial radius a in Eq. (1), much of recent work on Uranus' gravity field uses a' instead. We will conform to this usage in this paper, so that all calculated and observed J_{2n} are normalized to a' = 26,200 km.

The J_{2n} provide integral constraints on the structure of Uranus via the following relationships:

$$J_{2n} = \Lambda_{2n}q^n + \Lambda'_{2n}q^{n+1} + \ldots, \tag{2}$$

where

$$q = \omega^2 a^3/GM \tag{3}$$

is a dimensionless parameter and ω is the angular rotation velocity of the planet. The dimensionless response coefficients Λ_{2n}, together with the higher-order corrections $\Lambda'_{2n}, \ldots$ are functions of the planetary interior structure and serve as additional integral constraints (Zharkov and Trubitsyn 1978). For consistency in the following we will replace a with a' in Eq. (3), so that for Uranus and a rotation period of 15.5 hours we have q = 0.03935. Without knowledge of q, the J_{2n} themselves provide little or no information about interior structure.

From an analysis of the precession of the ε-ring, Nicholson, et al. (1978) found J_2 = 3.4 x 10^{-3}. This result was made more precise by Elliot, et al. (1981) who found J_2 = (3.354 $\pm$ 0.005) x 10^{-3}. They also obtained J_4 = (-2.9 $\pm$ 1.3) x 10^{-5}.

Some disagreement still exists about the correct rotation period for Uranus. Brown and Goody (1977) obtained 15.6 hours, Trauger, et al. (1978) obtained 13.0 hours, Elliot, et al. (1981) obtained 15.5 hours, and Franklin, et al. (1980) obtained 16.6

hours. On the other hand, Trafton (1977) found 23 hours and Hayes and Belton (1977) found 24 hours. Our preliminary conclusion, based on considerations presented below, is that more plausible interior models are associated with a rotation period of 15 hours, but that a 24-hour period cannot be absolutely ruled out.

If the density of the H_2-He atmosphere is increased by enrichment of other constituents beyond their solar proportions, this density increase can become gravitationally significant. Therefore it is important to obtain bounds on such possible enhancements. A thorough discussion of this problem is given by Wallace (1980). From an analysis of Uranus spectra at wavelengths from the visible to the microwave region, he concludes that the number ratio of CH_4 to H_2 in the deep atmosphere is greater than 0.01 and probably less than 0.10. If CH_4 is the only species enhanced above solar abundance, then these limits correspond to 0.06 $< f_{CH_4} < 0.4$, where f_{CH_4} is the ratio of the mass of CH_4 in the deep atmosphere to the total mass.

III. VARIANTS ON THE HM MODEL

As initiated by Podolak (1976), it is useful to plot the optical oblateness ε versus J_2 for various interior models. Here ε is the difference between the equatorial and polar radii in units of the equatorial radius. We have, to lowest order in q,

$$\varepsilon = q(3\Lambda_2 + 1)/2, \qquad\qquad\qquad (4)$$

so that a given interior model plots as a straight line with slope $(3 + \Lambda_2^{-1})/2$. When allowance is made for higher-order terms in Eq. (2), the lines are actually curved, but the curvature is negligible to the present order of accuracy.

Fig. 1 shows a plot of several interior models together with rotation periods of 15.5 hours and 24.0 hours. The three-layer H+M model is taken from HM (1980), while the Z+T model is taken from Zharkov and Trubitsyn (1978). It is similar to the H+M model except that it has only two layers: the "ice" and "rock" are homogeneously mixed in the interior below the H_2-He atmosphere. Also, the Z+T model uses an older H_2O equation of state which is significantly "harder" (i.e., has a higher pressure at a given density) than the one which we now use and which is favored by shock data on H_2O (Mitchell and Nellis 1979, Ree 1976). The dramatic effect of this revision in the H_2O equation of state is shown by the line marked "2-layer" in Fig. 1. This model is essentially identical to the Z+T model except that the new H_2O equation of state is used. Since Uranus is about one-half water, there is a major change due to the revision; the water is more compressed, the model is therefore more centrally condensed, and Λ_2 is smaller. For comparison, we have plotted observational data points for J_2 and ε from Elliot, <u>et al.</u> (1981) and Franklin, <u>et al.</u> (1980). These data seem to favor the revised "2-layer" model, i.e., the Z+T model with the updated H_2O equation of state. However, for reasons given above, this model

Fig. 1.   Optical oblateness ε as a function of J_2 for various Uranus models.

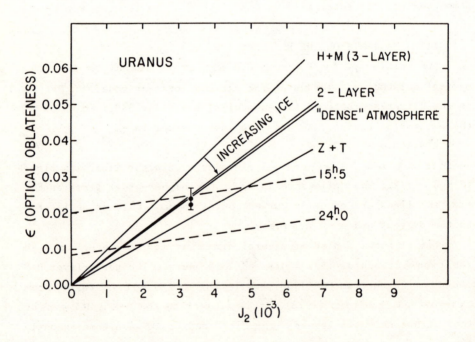

seems cosmogonically less plausible since it assumes that the "rock" and "ice" can remain uniformly mixed. Therefore it is important to note that this model does not give a unique fit to the data. We can also bring the H+M three-layer model into agreement by increasing I/R, as shown. This has the effect of increasing the moment of inertia because of the smaller rock core and smaller H_2-He envelope and therefore increasing Λ_2. One needs I/R $\sim$ 6 in order to bring the three-layer model into agreement with the data points. Since this exceeds the cosmic I/R ratio, what has become of the missing "rock" during formation of the planet?

The most cosmogonically plausible way to adapt the three-layer model to fit the data points is to assume that the H_2-He envelope contains a non-negligible amount of denser material such as CH_4. Increasing the density of the deep atmosphere has a significant effect on the moment of inertia, leading to a larger Λ_2. This requires $f_{CH_4} \sim 0.4$ to 0.5.

Is there any way to distinguish between the three models outlined above? For this purpose, we have computed the higher-order response coefficient Λ_4 using techniques which we will now describe.

IV. HIGHER-ORDER GRAVITY FIELD OF URANUS

Current data are not yet sufficiently accurate to permit Uranus models to be well-constrained by the higher J_{2n} (n > 1). Nevertheless, there are prospects for improvement based on study of the motions of the Uranian rings and an eventual flyby by the Voyager 2 spacecraft. Therefore we have begun a study of the higher gravity components of Uranus models with the goal of determining how well these can discriminate between different models with essentially identical Λ_2's.

Our calculation makes use of an approach which is described by Hubbard, et al. (1975), Slattery (1977), and Hubbard, et al. (1980). In this approach, we carry out a multipole expansion of the mass density $\rho(r,\theta)$ in the form

$$\rho = \sum_{n=0}^{\infty} \rho_{2n}(r) P_{2n}(\cos \theta) \qquad (4)$$

over all spherical regions of the planet which do not contain
discontinuities of ρ or singular points where derivatives of ρ do
not exist.   In Fig. 2, which schematically shows a rotating
two-layer planet, expansion (5) may be used in the innermost region
enclosed by a dashed line and it may also be used in the region
between the outermost dashed line and the next dashed line.   In the
spherical region which contains the oblate surface of discontinuity
between the core and the mantle, we use a set of shell integrals
(defined by Hubbard, et al., 1975) to determine the contribution to
the various Λ_{2n} from this region.   In the region where expansion (5)
is valid, the contribution to each Λ_{2n} is produced only by the
corresponding $\rho_{2n}(r)$.

     Fig. 2.   Division of a rotating, oblate planet into various
               interior zones for calculation of gravitational
               harmonics.

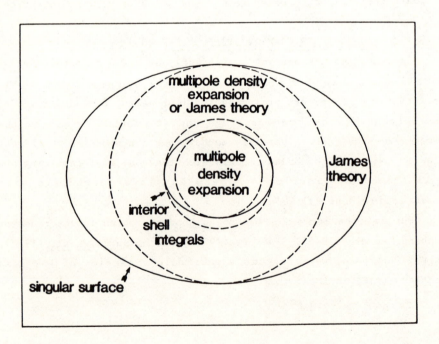

The surface (the 1-bar surface, say) requires special treatment. For an adiabatic variation of temperature and near-solar composition, the pressure-density relation resembles a polytrope of index 2.5 as the density approaches zero. As is well known, various derivatives of the density with radius become infinite as $\rho \to 0$, and thus it is not possible to accurately locate the surface by means of a Maclaurin series expansion from any interior point. To avoid this problem, we place a grid of quadrature points in r and θ in the outer region of the planet which includes the $\rho = 0$ surface. This approach was employed by James (1964) and does not make use of any assumed analyticity of the density distribution. For Uranus, our theory typically yields J_2 to an accuracy of $\sim 10^{-2}$%, and J_4 to ~ 0.4%. Values for J_6 and J_8 are also calculated but currently play no role in constraining interior models.

Most interior models of Uranus were calculated for a rotation period of $15^h.5$. For each model we then calculated effective response coefficients by the relations

$$\Lambda_2 = J_2/q, \tag{6}$$
$$\Lambda_4 = -J_4/q^2. \tag{7}$$

Comparing with Eq. (2), we see that this result is valid in the limit $q \to 0$. For finite q, Eqs. (6) and (7) thus include a very weak q-dependence in Λ_2 and Λ_4, but for our purposes this is negligible. The advantage of this representation for models is that the response coefficients are essentially independent of the adopted rotation period. Fig. 3 shows a plot of Λ_4 versus Λ_2 for a variety of Uranus interior models; all of these models were described above. The locus of points between the dashed lines is consistent with the results of Elliot, et al. (1981); their preferred rotation period is $15^h.5$. If the adopted rotation period is allowed to change, the observational constraints move as shown while the model values remain essentially fixed.

On the left we have three sequences of models of the H+M type, i.e., with "rock" cores, "ice" envelopes, and H_2-He atmospheres. For any given value of I/R, there is a range of models produced by varying the thickness of the H_2-He atmosphere. The range shown corresponds to a reasonable range in a about the observed value of 25,900 km.

The models indicated with short dashed lines are of the two-layer Z+T type, i.e., with H_2-He atmospheres and homogeneous "ice"-"rock" interiors. The updated H_2O equation of state is used, however.

Interestingly, the addition of substantial amounts of methane to the H_2-He atmosphere of an H+M model (models with heavier lines) can produce a substantial decrease in Λ_4 for fixed Λ_2. Since models of this type are the most cosmogonically plausible, it is interesting to see that they are also consistent with the observations. Note that Wallace sets a limit $f_{CH_4} < 0.4$ for the deep Uranus atmosphere. There is still a great variety of possible models.

Fig. 3.   The response coefficient Λ_4 as a function of Λ_2 for various Uranus models. Current observational constraints on the gravity field limit models to the region between the dashed lines.

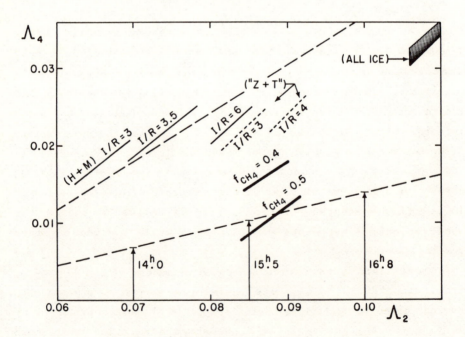

Finally, we have considered the possibility of a Uranus model with a solid or liquid "surface" that is directly observationally accessible. Such a model would require an extremely thin hydrogen-rich atmosphere, extending only up to pressures $\sim$50 bar. Such an atmosphere would be gravitationally negligible, and the planet would basically have to consist of "rock" and "ice" alone. With this composition, the model turns out to be too dense; the best fit to the observed mean density is achieved with pure "ice" and no "rock" core at all, but such a model still ends up with a radius a $\approx$ 23,000 km, i.e., about 10-15% too small. One may argue that the equation of state of the "ice" component is sufficiently uncertain that such a model cannot be excluded. Allowing for such uncertainties, we estimate that suitable "all-ice" models would be found in the location indicated in Fig. 3.

V.   CONCLUSIONS

Recent observations of parameters relevant to Uranus' interior are beginning to point toward a particular type of interior model. This model has a "rock" core, an "ice" layer, and an H_2-He atmosphere with copious amounts ($f_{CH_4} \sim 0.4$) of methane. The overall value of I/R is about 2.5. Apparently methane is soluble in the atmosphere although ammonia may not be because it is observed to be depleted (Wallace, 1980). Perhaps there is some stirring between the atmosphere and the "ice" layer since molecular diffusion times are very long. It is difficult to think of an alternative, cosmogonic way of producing the large methane enrichment.

On the other hand, before such a model is too enthusiastically adopted, we should mention two troublesome observational constraints. The first comes from spectroscopic studies of Uranus' atmosphere, which show that the deuterium to hydrogen ratio (D/H) there is approximately the same as that in Jupiter and Saturn and approximately primordial (Hubbard and MacFarlane, 1980). When methane or water condenses from a primordial solar nebula, temperatures are low enough to cause substantial fractionation of deuterium into H_2O and CH_4. Thus we would expect D/H to be at least four times larger than the primordial ratio in Uranus' atmosphere. The observations seem to imply that Uranus' atmosphere is not stirred with the interior. But in this case, why is methane so enriched in the atmosphere?

The second problem arises from the fact that there is evidence that CH_4 tends to dissociate at pressures above 200 kbar and temperatures above 2000 K, i.e., probable conditions within the "ice" layer (Ross and Ree, 1980). Apparently the methane decomposes into hydrogen and elemental carbon. If the carbon sinks and the hydrogen rises, this would tend to deplete methane in the atmosphere, contrary to observation. Therefore the decomposition of methane, if it occurs, either plays no role because there is no communication between the atmosphere and the interior, or else it occurs in a highly reversible manner.

It should also be pointed out that the interior temperature profile of Uranus (in an adiabatic model) can be significantly altered if H_2O is significantly enhanced in the deep atmosphere as well as CH_4. Calculations by Hunten (1981) indicate that internal temperatures are reduced by $\sim 10\%$ if H_2O has a number abundance of 3% relative to H_2. This effect has not been included in models discussed here.

As discussed by HM and by Hubbard (1978), the structure of the Uranian atmosphere may play an important role in the heat balance of Uranus as well. A measurement of the actual intrinsic heat flow rather than the upper limit which is currently available, may play an important role in helping to understand the cosmogony of Uranus and its present fluid circulation patterns.

ACKNOWLEDGEMENTS

We thank D. M. Hunten, J. L. Elliot, and M. Ross for advance copies of papers relevant to this discussion. This research was supported by NASA Grant NSG-7045.

REFERENCES

Brown, R. A., and Goody, R. M. (1977). The rotation of Uranus. Astrophys. J. 217, 680–687.

Danielson, R. E., Tomasko. M. G., and Savage, B. D. (1972). High resolution imagery of Uranus obtained by Stratoscope II. Astrophys. J. 178, 887–900.

Elliot, J. L., French, R. G., Frogel, J. A., Elias, J. H., Mink, D., and Liller, W. (1981). Orbits of nine Uranian rings. Astron. J., 86, 444–455.

Franklin, F. A., Avis, C. C., Colombo, G., and Shapiro, I. I. (1980). The geometric oblateness of Uranus. Astrophys. J. 236, 1031–1034.

Hayes, S. H., and Belton, M. J. S. (1977). The rotation periods of Uranus and Neptune. Icarus 32, 383–401.

Hubbard, W. B. (1978). Comparative thermal evolution of Uranus and Neptune. Icarus 35, 177–181.

Hubbard, W. B. (1981). Constraints on the origin and interior structure of the major planets. Phil. Trans. Roy. Soc., in press.

Hubbard, W. B., and MacFarlane, J. J. (1980). Structure and evolution of Uranus and Neptune. J. Geophys. Res. 85, 225–234.

Hubbard, W. B., and MacFarlane, J. J. (1980). Theoretical predictions of deuterium abundances in the Jovian planets. Icarus 41, in press.

Hubbard, W. B., MacFarlane, J. J., Anderson, J. D., Null, G. W., and Biller, E. D. (1980). Interior structure of Saturn inferred from Pioneer 11 gravity data. J. Geophys. Res. 85, 5909–5916.

Hubbard, W. B., Slattery, W. L., and DeVito, C. L. (1975). High zonal harmonics of rapidly rotating planets. Astrophys. J. 199, 504–516.

Hunten, D. (1981). In preparation.

James, R. A. (1964). The structure and stability of rotating gas masses. Astrophys. J. 140, 552–582.

Mitchell, A. C., and Nellis, W. J. (1979). Water Hugoniot measurement in the range 30-220 GPa. High Pressure Sci. Technol. 1, 428-434.

Mizuno, H. (1980). Formation of the giant planets. Progr. Theoret. Phys. 64, 544-557.

Nicholson, P. D., Persson, S. E., Matthews, K., Goldreich, P., and Neugebauer, G. (1978). The rings of Uranus: results from the 10 April 1978 occultation. Astron. J. 83, 1240-1248.

Podolak, M. (1976). Methane rich models of Uranus. Icarus 27, 473-477.

Podolak, M., and Cameron, A. G. W. (1974). Models of the giant planets. Icarus 22, 123-148.

Podolak, M., and Reynolds, R. T. (1981). On the structure and composition of Uranus and Neptune. Icarus, 46, 40-50.

Ree, F. (1976). Equation of state of water. Rep. UCRL-52190, Lawrence Livermore Lab., Livermore, Calif.

Reynolds, R. T., and Summers, A. L. (1965). Models of Uranus and Neptune. J. Geophys. Res. 70, 199-208.

Ross, M., and Ree, F. H. (1980). Repulsive forces of simple molecules and mixtures at high density and temperature. J. Chem. Phys. 73, 6146-6152.

Slattery, W. L. (1977). The structure of the planets Jupiter and Saturn. Icarus 32, 58-72.

Trafton, L. (1977). Uranus' rotational period. Icarus 32, 402-412.

Trauger, J. T., Roesler, F. L., and Munch, G. (1978). A redetermination of the Uranus rotation period. Astrophys. J. 219, 1079-1083.

Wallace, L. (1980). The structure of the Uranus atmosphere. Icarus 43, 231-259.

Zharkov, V. N., and Trubitsyn, V. P. (1972). Adiabatic temperatures in Uranus and Neptune. Izv. Akad. Nauk SSSR Fiz. Zemli 7, 120-127.

Zharkov, V. N., and Trubitsyn, V. P. (1978). Physics of Planetary Interiors (edited and translated by W. B. Hubbard), Pachart, Tucson.

THE MAGNETOSPHERE OF URANUS

W.I. Axford

Max-Planck-Institut für Aeronomie, D-3411 Katlenburg-Lindau 3

I. THE NATURE OF PLANETARY MAGNETOSPHERES

The magnetosphere of a planet, as the name implies, is the region surrounding the planet in which the planetary magnetic field plays a dominant role in determining the behaviour of the medium[1]. The inner boundary of a magnetosphere is the surface of the planet if it has no significant atmosphere (as in the case of Mercury), or the lower ionosphere in the case of planets with atmospheres. The outer boundary, usually termed the "magnetopause", is shaped by stresses exerted by the solar wind, being blunt on the upstream side, with normal stresses playing a dominant role, and extending in a comet-like "magnetotail" on the downstream side away from the Sun largely as a result of the action of shear stresses. The characteristic size of a magnetosphere, namely the distance L_m (in planetary radii) to the subsolar point is determined approximately by balancing solar wind ram pressure and the magnetic pressure:

$$nmV_s^2 = C\ B_o^2/2\pi L_m^6 \tag{1}$$

where n is the ion number density and V_s the speed of the solar wind, m is the average ion mass, B_o the (dipole) field strength at the surface of the planet and C is a constant of order unity.

In general, the magnetic pressure in a magnetosphere is much greater than that of the plasma contained and hence the magnetic field configuration is approximately that of a potential field with the given boundary conditions. There are exceptions, however, such as in the case of Jupiter, where the pressure of magnetospheric plasma and rotational stresses play a very important role in determining the magnetic field configuration. These additional stresses produce distributed currents within the magnetosphere, equivalent, for example, to an equatorial disk current distribution for the case of rotational stresses. The plasma pressure must always be dominant in the strong current sheets which appear in the magnetotail separating regions of out-going and in-going magnetic flux.

In all cases planetary magnetospheres are sufficiently large obstacles in the supersonic solar wind that bow shock waves must occur on the upstream side ahead of the magnetosphere. The chief function of the shock wave is to reduce the solar wind to subsonic speeds so that the flow may take account of the presence of the magnetosphere and flow around it. The shock wave itself, being collisionless, produces plasma wave turbulence, plasma heating and, in suitable circumstances, acceleration of particles to energies very much greater than that of the incident solar wind ions (i.e. >> 1 keV/nucleon)[2].

The interplanetary magnetic field carried by the solar wind plays an important role in the interaction with the magnetosphere in that it can reconnect with the magnetospheric field at the magnetopause and so change the topology and linkage of field lines (defined in the usual way by the plasma "frozen" to them). This has the effect of converting "closed" magnetic field lines (i.e. connected at each end to the planet itself) into "open" field lines (connected to the planet at only one end and leading into the interplanetary medium). Energetic particles can be trapped in closed regions of the magnetosphere, forming radiation belts for example, while open magnetic field lines permit the free access of particles of external origin and the free escape of magnetospheric particles[3-5].

Open magnetic field lines must eventually reconnect with each other to form closed field lines once again. This process, which takes place in the magnetotail, is likely to be the most important one for producing hot plasma and energetic particles which can subsequently be injected into the closed inner magnetosphere[6]. The upstream and downstream reconnection processes are in balance only on the average so that the ratio of closed to open magnetic flux can vary with time. Furthermore, there are reasons to believe that reconnection occurring in the magnetotail can be spontaneously and sporadically enhanced producing the phenomenon which is usually called a magnetospheric "substorm" and which has very close simila-rities to the solar flare phenomenon[7,8]. Magnetospheric "storms" are the consequence of major changes in the solar wind flow asso-ciated with solar disturbances such as flares and these are inevitab-ly accompanied by strong substorm activity, at least in the case of the Earth.

As a direct consequence of reconnection a component of the mag-netic field normal to the magnetopause (B_n) exists as well as a tan-gential component (B_t). Furthermore, the plasma at the magnetopause moves so that the Maxwell shear stress $B_n B_t / 4\pi$ produces an energy exchange at a rate

$$W = \frac{1}{4\pi} \int_A B_n B_t V_t \, dA \tag{2}$$

where V_t is the plasma speed and the integration is carried out over the magnetopause surface (A). On the forward side of the magnetopause the sense of the stress is such that the magnetic field energizes the plasma at a rate

$$W_{m1} = \frac{1}{4\pi} \int_A (V_t B_n) B_t \, dA = \frac{1}{4\pi} \int_A E_m B_t \, dA \sim \frac{1}{8} \, \phi B_m L_m \tag{3}$$

where E_m is the electric field and B_m the magnetic field at the mag-netopause, and ϕ the corresponding potential drop. On the tailward side of the magnetopause the external solar wind plasma works against the Maxwell shear stress and so builds up the magnetic energy stored in the magnetotail at a rate

$$W_{m2} = \frac{1}{4\pi} \int_A V B_t (B_n dA) \sim \frac{1}{4\pi} \, V B_t F \tag{4}$$

where F is the total open magnetic flux and B_t the field strength in the magnetotail. Note that B_t is determined (at large distances from the planet) by the pressure of the external interplanetary medium (excluding the ram pressure) and hence $B_t \sim B_m / M$, where M is the solar wind Mach number ($M \sim 10$ everywhere). Finally, as a conse-

quence of reconnection occurring at the 'neutral' sheet in the magnetotail the average rate of conversion of magnetic to plasma energy
is

$$W_t = \frac{1}{4\pi} \int_A B_t^2 V_n dA \sim \frac{B_t^2 V_n A_t}{4\pi} \sim \frac{\phi B_t L_t}{4\pi} \tag{5}$$

where A_t is the area of the neutral sheet under consideration, L_t its
length and V_n is the plasma speed towards the sheet.

II.   ELECTRIC FIELDS IN MAGNETOSPHERES

An essential feature of magnetospheres which has an important
bearing on the behaviour of the plasma and energetic particles they
contain, is that large scale electric fields exist[9]. There are
corresponding motions of the plasma such that in general the "frozen
field" condition $\underline{E} + \underline{V} \times \underline{B} \sim 0$ is satisfied (except in restricted regions with strong electric currents flowing parallel to the field
lines where a significant $E_{\shortparallel}$ can arise). These hydromagnetic motions
of the plasma within a magnetosphere are generally called "convection" in the sense that one imagines the magnetospheric magnetic
field lines to be moving and carrying the plasma along with them. The
motions may be quasi-steady such as those induced by planetary rotation, atmospheric tides and the average solar wind/magnetopause
interaction, or unsteady, such as those induced by variations in
solar wind conditions, substorms and low frequency Alfvén and mag-
netoacoustic waves.

The convection electric fields provide a means for accelerating
and decelerating the magnetospheric plasma and energetic particles
as well as transporting plasma into and out of the magnetosphere.
The energization can be regarded as an adiabatic effect associated
with changes in the volume of magnetic flux tubes as they move radi-
ally and change shape. In the case of quasi-steady convection the
energy changes are limited by the potential differences associated
with the electric field pattern (typically $\sim$ 10% of the external potential difference). In the case of time-varying and random convection, however, there is no such constraint because the electric field
is not steady and a few particles will always be able to move from
regions of low magnetic field strength to regions of high magnetic

field strength ("radial diffusion") with a corresponding increase of energy[10].

It is important to note that convective motions are to some extent controlled by the pressure gradients of the magnetospheric plasma and energetic particles as well as by the driving mechanisms mentioned above[11]. The pressure gradients can be regarded as making their presence felt as a result of particle drifts associated with inhomogeneities of the plasma and magnetic field which tend to set up space charge distributions and therefore additional electric fields. This effect can be partly but not completely neutralized by currents flowing to and from the ionosphere along magnetic field lines which in turn may cause parallel electric fields to arise if the density of the plasma required to carry the current is sufficiently low. The parallel electric fields accelerate particles into and out of the ionosphere, thus giving rise to auroral effects when the particles strike the upper atmosphere and possibly introducing new particles of ionospheric origin into the outer magnetosphere[12].

Electric fields are of particular importance in the vicinity of regions where magnetic field line reconnection takes place (loosely described as "neutral" or "X" points). Such electric fields may become transiently and locally very large in cases where the reconnection is sporadic and short-lived. The speed of reconnection (and therefore the electric field and voltage drop available) is limited only by the Alfvén speed (V_A) in the medium so that large transient voltage drops ($\sim V_A B L_m$) are possible as long as the magnetic field configuration is favourably configured for reconnection to occur, as seems to be the case during substorms. In the Earth's magnetosphere, for example, it is apparently possible for voltage drops of the order of ~ 1 MV to exist in the magnetotail for times of the order of $\sim 10-10^2$ sec[13], which is very much larger than the voltage drops associated with other forms of convection (e.g. rotation ~ 88 kV, solar wind driven convection $\sim 50-200$ kV, tidally driven convection $\sim 30-50$ kV). The voltage drops associated with reconnection are especially interesting because it is always possible for a few par-

ticles to move a considerable distance along the neutral line, consequently being rapidly accelerated to comparatively high energies.

III. SOURCES OF PLASMA AND ENERGETIC PARTICLES IN MAGNETO-
SPHERES

The chief sources of plasma in planetary magnetospheres are the
solar wind and ionospheres of the planet and any satellites contained in its magnetosphere. In addition, magnetospheres are usually pervaded by a neutral component consisting of the exosphere of
the planet (comprised largely of hydrogen), the exospheres of satellites, sputtered products from satellite surfaces and dust and the
neutral interstellar medium (which penetrates to within 4 AU of the
Sun in the case of hydrogen and 0.5 AU in the case of helium[14]).
In addition, there are albedo neutrons and mesons produced as a result of the interaction between high energy cosmic rays and the planetary and satellite atmospheres and/or surfaces and also planetary
rings.

Neutron decay can be an important source of energetic protons
and electrons in regions of the magnetosphere close to the planet.
However, the fluxes of neutrons are very low and it is necessary
that the life times of the trapped particles be very long ($\sim$ years)
in order to maintain a substantial radiation belt. The relevant loss
mechanisms are charge exchange with the neutral background (in the
case of protons) and pitch angle scattering followed by loss into
the atmosphere of the planet. The latter may be the result of collisions with background plasma and neutral gas and plasma wave scattering due to waves injected from below (lightning flashes or man-
made disturbances) or induced by plasma instabilities in the magnetosphere itself. A further important loss mechanism is absorption by
planetary dust rings and satellites; this is of particular interest
in regions where the energetic particle population is long-lived,
since in such cases, the absorption lanes produced by satellites,
for example, must be sharply defined.

Planetary ionospheres are a possible source of plasma, especially protons which tend to dominate at high altitudes and for which
gravitational binding is least effective. In general, if magneto-

spheric convection and reconnection with the interplanetary magnetic field permits the plasma to escape into space, the upper ionosphere should flow continually away from the planet in the form of a "polar wind" [15] driven by electron heat conduction or hydromagnetic wave pressure gradients. In regions of the magnetosphere where escape into space is not possible, the plasma pressure will build up until an approximate hydrostatic equilibrium is achieved between the upper magnetosphere and upper ionosphere. A region where such an equilibrium has been achieved is called a "plasmasphere" and its boundary ("plasmapause") is defined by the outermost closed flow line of the convection pattern (at least in the steady state). [16]

Satellite ionospheres can be a copious source of plasma in cases where the satellites have sufficiently dense atmospheres to produce an ionosphere (e.g. Io and Titan). In such cases gravity does not play an important role and the plasma tends to be carried away by the convecting planetary magnetic field lines provided these are not captured by the satellite and/or are not so much distorted that they become disconnected from the planetary magnetic field. Since gravity is not as important as in the case of the planetary ionosphere it is to be expected that heavier ions than protons can be injected into the magnetosphere in this way (e.g. S, O, and SO_2 ions in the case of Io[17]). The existence of a satellite magnetic field of internal origin can protect its ionosphere from such stripping of plasma by the planetary magnetosphere if it is sufficiently strong to produce a region of closed magnetic field lines.

Neutral gas may also escape from satellites either by evaporation of their atmospheres or sputtering from their surfaces. Escaping particles with low energies remain trapped in orbit around the planet forming a neutral gas torus (H in the case of Titan and Na, K and SO_2 products in the case of Io)[18]. Neutral gas tori are also sources of plasma following collisional photo-ionization and also act as a loss mechanism for magnetospheric ions as a result of charge exchange. The process of ion mass pick-up involved when satellite upper atmosphere and torus atoms and molecules become ionized is interesting in that the new ions are accelerated to energies $T_\perp$ =

$\frac{1}{2} mV_r^2$, where V_r is the convection speed relative to the satellite or neutral gas cloud. Furthermore, a pick-up electric current I is induced such that

$$V_r \dot{M} = I B L_n \tag{6}$$

where $\dot{M}$ is the total mass ionization rate and L_n is the characteristic dimension of the neutral gas cloud or atmosphere[19]. The current distorts the magnetic field pattern and must close through the planetary ionosphere allowing some possibility for the generation of radio wave emissions and parallel electric fields with associated particle acceleration[20].

It is in principle possible for plasma particles originating from the planet and satellites to be accelerated to high energies by interacting stochastically with plasma wave turbulence[10]. This is not a very efficient process in general, however, because there are no very powerful external sources of plasma turbulence and the waves are not always confined within the magnetosphere. The most effective acceleration mechanism for such particles is likely to be reconnection occurring on stretched-out field lines in the magnetotail followed by convection and/or diffusive transport back into the inner magnetosphere. In such a scheme there are no difficulties with the total energy requirement since the magnetic energy stored in the magnetotail can be tapped and this is in turn drawn from the solar wind and possibly in some cases from the rotational energy of the planet. Direct acceleration from ionospheres is possible as a result of the occurrence of parallel electric fields and if accompanied by energetic particle precipitation these may result in rather peculiar ions being introduced into the magnetospheric energetic particle population (e.g. H_2^+ and H_3^+ in the case of Jupiter and Saturn[21]).

The solar wind must always be considered as a possible source of plasma for a planetary magnetosphere. The mechanisms of entry include drift and diffusion across the magnetopause and flow along open field lines produced by reconnection between the interplanetary magnetospheric magnetic fields. Final capture of solar wind plasma on open field lines occurs only after reconnection occurs in the tail forming closed field lines which can convect the plasma into the inner magnetosphere. Since the plasma on open magnetotail field lines, whether of solar wind or planetary origin, tends to stream away from the Sun it is necessary for efficient capture of solar wind plasma that tail reconnection takes place at large distances from the planet. In the case of the Earth's magnetosphere, where plasma of internal origin does not play an important role in determining the dynamics of the situation, the capture efficiency (ϵ) for solar wind plasma is of the order of 10^{-3}-10^{-4} [22]. Clearly, the solar wind plasma has a special signature in that the relative abundances of elements and isotopes correspond to those of the solar corona and the ionization state is that of a plasma with an electron temperature of the order of 1.5×10^6K. Thus one expects to find in addition to protons, ions such as He^{+2} and O^{+6}, whereas particles of internal origin have "non-solar" elemental abundances and tend to be singly ionized unless the magnetospheric plasma is sufficiently dense and hot as to permit ionization by electron impact to higher levels.

The interstellar gas is a further possible source of magnetospheric particles, which is important provided it can be ionized with sufficiently high efficiency. The ratio of the interstellar and solar wind sources is given approximately by

$$\alpha N_n L_m / 3 \epsilon \Phi_{sw} \tag{7}$$

where α is the ionization rate, N_n the density of the interstellar medium, L_m the characteristic size of the magnetosphere and Φ_{sw} the solar wind flux. If photo-ionization of the interstellar hydrogen is the dominant ionization process then both α and Φ_{sw} scale similarly with distance from the Sun and, in particular, $\alpha \sim 1.5 \times 10^{-7}$/sec and $\Phi_{sw} \sim 2 \times 10^8$/cm² sec at 1 AU. Taking $N_n \sim 10^{-1}$/cm³ and $\epsilon \sim 10^{-4}$, we see that the interstellar source is not likely to be important unless $L_m \gtrsim 10^{12}$cm, which is much larger than found for any magnetosphere observed to date.

IV. COMPARISON OF MAGNETOSPHERES

To date, magnetospheres have been found around the Earth, Mercury, Jupiter and Saturn (in order of discovery). Venus appears to have no significant magnetic field of internal origin, as do the comets, hence the description of a magnetosphere given above is not useful. Nevertheless, the term is sometimes applied to the region of strong magnetic field captured by the ionosphere from the interplanetary medium in such cases. It seems possible that a very weak magnetic field of internal origin exists in the case of Mars and hence one may consider that this planet has a genuine magnetosphere, at least transiently, but it is very much on the border line.

The Earth's magnetosphere has by now been rather thoroughly explored by a large number of spacecraft, although by no means everything is well understood. The strength of the Earth's magnetic dipole (0.3 gauss at the equator) is such that the characteristic dimension of the magnetosphere on the upstream side is typically ~ 10 R_e (Earth radii), although variations from ~ 5 R_e to 12 R_e are possible. Rotation and atmospheric tidal effects play a relatively minor role and most of the energy input is provided by the solar wind. From (4) we deduce that, with $B_t \sim 10$–20 gauss, about 1% of the incident solar wind energy flux (10^{20}–10^{21}ergs/sec) goes into forming the magnetotail (i.e. 10^{18}–10^{19}ergs/sec) and about 10% of this ultimately appears in the inner magnetosphere in the form of energetic particles, some of which are lost into the atmosphere producing aurorae. The total energy stored in the magnetosphere is $\sim$ 10^{20}–10^{22}ergs and the average dissipation rate about 10^{17}–10^{18}ergs/sec[23]. Most of the energetic particles in the magnetosphere appear to originate in the solar wind[24] but a substantial component (10% or more) of ionospheric origin is sometimes observed[25]. The contribution from atmospheric neutron albedo is very small. Since the Moon's orbit crosses only the distant magnetotail it has no significant influence on the magnetosphere, although interesting absorption effects can be seen in its immediate vicinity[26]. The inner boundary of solar wind induced convection lies usually in the range 3–6 R_e, depending on magnetic activity, and within this region a rather dense plasmasphere is formed (densities 10^2–10^3cm^{-3} at the plasmapause) comprised largely of H^+ together with smaller amounts of He^+ and O^+ ions[27].

The undistorted equatorial field intensity of Mercury appears to be of the order of 260 gammas[27], which yields a minimum radius of the magnetopause of the order of 1.5 R_m. On the basis of terrestrial values, one might expect an energy influx of $\sim 10^{16}$-10^{17} ergs/ sec to the magnetotail and $\sim 10^{15}$-10^{16} ergs/sec to the inner magnetosphere, with a solar wind particle injection rate of 10^{22}-10^{23}/sec. Despite the relatively small size of the magnetosphere the low plasma density and relatively high magnetic field strength ($\sim$ 50-100 gammas) in the magnetotail allow large transient electric fields to develop and accordingly particles are observed to be accelerated to quite high energies[28]. In view of the fact that the planet itself occupies a large part of the magnetosphere by volume, most of the particles trapped by the solar wind are presumably absorbed directly on the surface which should produce a continuous flux of sputtered ions and neutral atoms which may be observable spectroscopically[29]. Mercury has no significant atmosphere or ionosphere and hence magnetospheric convection is impeded only by plasma pressure effects.

The magnetosphere of Jupiter is characterized by its very large size (4 gauss equatorial surface field and $\sim$ 50-100 R_J minimum radius), and the presence of the satellite Io in the inner magnetosphere. Io has a thin atmosphere associated with venting of gases such as SO_2 and an ionosphere which contributes a large flux of ions to the magnetosphere ($\sim 10^{28}$/sec)[17,30]. The mass pickup current, according to equation (4), is $\sim 10^6$ amps and the corresponding energy dissipation $\sim 10^{19}$ ergs/sec. As a consequence of the rapid rotation of the planet, which dominates the convection, the centrifugal stresses are sufficient to distort the configuration of magnetospheric field lines into a disc-like structure which extends into the magnetotail forming a plasma sheet superficially similar to that of the Earth[31]. There is a distinction, however, in that at Jupiter the plasma in this sheet is largely of internal magnetospheric origin and escapes down the tail whereas in the Earth's magnetosheath the plasma tends to be pre-dominantly of solar wind origin and is brought from the tail into the inner magnetosphere.

There is evidence for the presence of particles of solar wind origin
at higher energies in the Jovian magnetosphere and also, surprisingly enough, particles of ionospheric origin (notably H_2^+ and H_3^+),
presumably being extracted from the planetary ionosphere in some
form of auroral process[21]. The total energy flux involved in these
processes is $\sim 10^{22}$ergs/sec with the rotational energy of the planet
being probably the most important source. The particle injection
rate is of the order of 10^{28}/sec with Io, the planetary ionosphere
and the solar wind being of rather comparable importance[32]. Presumably the effects of rotation drive an unstable convection
pattern[33-35] giving rise to radial diffusion of energetic particles. The latter may be given a significant boost in energy as a
result of reconnection taking place in the magnetotail.

The Saturnian magnetosphere is rather more similar to that of
the Earth than of Jupiter, despite its rapid rotation rate. The
chief reasons for this are that the Saturnian magnetic field is relatively weak (~ 0.2 gauss at the equator)[36] and there is no
strong source of plasma within the magnetosphere comparable to Io.
The satellite Titan, which has a dense atmosphere and ionosphere,
has an important effect on the outer magnetosphere as it produces in
particular a neutral hydrogen torus which provides an upper limit to
the lifetime of protons of $\sim 10^7$ sec[37]. The energetic particles in
the magnetosphere appear to be largely of solar wind origin apart
from a component comprised of molecular hydrogen ions which must
originate in the ionospheres of Titan or Saturn itself[38]. There is
evidence for a satellite source of plasma since large numbers of
oxygen ions are present in the inner magnetosphere where there are
several moderate sized satellites with icy surfaces[39]. By analogy
with the terrestrial magnetosphere the solar wind energy input to
the Saturnian magnetosphere is of the order of 10^{19}-10^{20}ergs/sec.
The presence of the rings and a number of small satellites close to
the planet is, however, of some interest since cosmic ray interactions producing neutron albedo and mesons appear to be the dominant source of the more energetic particles and the absorption lanes
produced by satellites are very sharply defined indicating that there
is little if any radial diffusion[40-42] in the inner magnetosphere.

V. PROSPECTS FOR URANUS

At present the literature devoted to the magnetosphere of Uranus is very sparse indeed, consisting of a report of a possible observation of radio emissions from the planet[43], a number of discussions contained in proposals for the ill-fated MJU mission[44,45] and a few papers speculating on the possible nature and magnetic topology of the Uranian magnetosphere[46-48].

It is generally agreed that since in all other cases so far observed, the magnetic dipole axis and the axis of rotation of the planet are approximately parallel this should also be the case at Uranus. Since at the present time the axis of rotation is pointing almost directly at the Sun this implies that the Uranian magnetosphere is unusual in that the dipole axis is roughly parallel rather than roughly perpendicular to the direction of the solar wind flow. Furthermore, one polar region is constantly sunlit and thus has a permanent ionosphere with an electron density typically of the order of 10^4cm^{-3}, whereas the other pole is in darkness and any ionosphere that may exist can only be due to magnetospheric particle precipitation and a very weak contribution from galactic radiation. The rate of rotation of Uranus is somewhat longer than that of Jupiter and Saturn and it is also noteworthy that, although five small (probably icy) satellites and particulate rings are present, there is no obvious strong source of plasma and neutral gas such as Io and Titan.

The radius of Uranus is only half that of Saturn and its rate of rotation also slower, however, the angular momentum of the planet is comparable to that of both Saturn and Jupiter. In view of our ignorance of the internal structures of the planets and the lack of a quantitative magnetic dynamo theory it is not reasonable to do anything more than guess that the surface field of Uranus might lie in the range $B_o \sim 0.1\text{-}1.0$ gauss. Accordingly, the minimum radius of the magnetosphere, for average solar wind conditions at the orbit of Uranus, is given approximately by

$$L_u \sim 35 \, B_o^{1/3} \sim 25\text{-}50 \, R_u \qquad (8)$$

with $B_m \sim 5$ gammas. A magnetosphere of this size would usually en-

close the entire Uranian satellite system but as the satellites are
so small and probably lacking all but the most tenuous of atmospheres, one does not expect them to produce any significant magnetospheric effects other than acting as absorbers for energetic
particles and plasma and possibly as a source of sputtered neutrals,
which would probably be dominated by the products of water ice.

The bow shock standing ahead of the Uranian magnetosphere
should accelerate particles to suprathermal energies as observed at
the Earth and Jupiter. However, since the interplanetary magnetic
field is on the average almost perpendicular to the solar wind direction at this distance from the Sun, it is not expected that the
acceleration should be very efficient for particles originating in
the solar wind. Suprathermal particles of interplanetary and solar
origin should, however, be accelerated by the single reflection
mechanism, but probably not very effectively by diffusive (multiple)
reflection since the time scales available are relatively short[49].

It is to be expected that, as in the case of Saturn, the
Uranian magnetosphere is not an especially active one and the radiation belts should not be very intense. The solar wind energy flux in-
cident on the Uranian magnetosphere should be of the order of $2-8 \times 10^{19}$ ergs/sec under normal conditions, increasing by perhaps a factor
10 during disturbances. The rate at which energy is injected into
the magnetosphere in this way is therefore probably of the order of
$2-8 \times 10^{17-18}$ ergs/sec, which is somewhat less than that found at the
Earth, for example. The corresponding particle injection rate is of
the order of $10^{25}-10^{26}$/sec.

The most interesting aspect of the Uranian magnetosphere concerns the nature of its interaction with the solar wind, in view of
the fact that it is unusual in being pole-on to the flow. As a consequence of the normal stresses exerted by the solar wind plasma and
magnetic field it is to be expected that a distinct funnel is formed
on the upstream side of the magnetosphere. The funnel should be de-
flected to one side as a result of reconnection between the planetary magnetic field and the interplanetary magnetic field, which on
the average should lie in the ecliptic plane and be approximately
perpendicular to the Sun-planet line. As a consequence of reconnec-

tion solar wind particles should be able to penetrate easily into the region near the pole on the sunward side of the planet causing some additional airglow in this region.

Freshly opened magnetic field lines must, in the usual manner, be stretched out and add to the two halves of the magnetotail. The open field lines contained in the magnetotail should consist of a roughly circular bundle emanating from the magnetic pole on the dark side of the planet, separated by a thin plasma sheet from a more crescent-shaped bundle of field lines with the opposite field direction. The latter bundle may also be displaced out of the ecliptic plane in the sense of planetary rotation by an amount dependent on the balance of stresses exerted by the ionosphere at the foot of the bundle and by the solar wind on the interplanetary side. The magnetic field strength in the magnetotail should be of the order of 0.5 gamma for average solar wind conditions.

The plasma sheet separating the two regions of oppositely directed magnetic fields in the magnetotail must map into a closed curve in the polar regions on the dark side of the planet. This curve should mark the high "latitude" edge of the auroral zone on the dark side. The auroral zone may not extend more than a few degrees in latitude because the total electric potential drop available from planetary rotation may be of the order of 10^{6-7}V whereas the solar wind induced convection (assumed to be about 1/10 of the interplanetary potential drop across the magnetosphere) is only of the order of 5×10^4V. As a consequence, the magnetosphere could in principle contain a very large plasmasphere extending almost to the magnetopause. Since all moderately high latitude magnetic field lines have one end connected to a fully sunlit ionosphere a weak polar wind of protons and electrons should be sufficient to provide the necessary plasma if the escape time is long. It is conceivable that the distribution of plasma in the plasmasphere is thus dominated by rotational instability as in the case of Jupiter, but whatever the cause of radial diffusion it is likely to be relatively slow so that neutralization by the satellites could be a significant loss mechanism as well as escape along opened field lines into the magnetotail[34].

In such a quiescent magnetosphere one would expect the low

energy plasma to be dominated by protons of ionospheric origin and possibly ions such as O^+ sputtered from the surfaces of the satellites and rings. The more energetic particles in the outer magnetosphere could be comprised of accelerated plasmaspheric ions, some molecular hydrogen ions as found at Jupiter and Saturn and ions of solar wind origin. The most energetic particles are likely to be the result of interactions between cosmic rays and the satellites, rings and atmosphere of the planet but there may also be a contribution from temporarily trapped particles of interplanetary or solar flare origin. As a consequence, much of the Uranian magnetosphere should be rather similar to the inner parts of the Saturnian magnetosphere and relatively stable in comparison with the magnetospheres of the Earth and Jupiter.

VI. CONCLUSIONS

On the basis of the above arguments and guesswork, it appears that the Uranian magnetosphere should be characterized by long time scales in the inner regions and hence sharply-defined satellite and ring absorption lanes and relatively low energetic particle intensities. Nevertheless, considerable activity may occur in the outermost parts of the magnetosphere and the magnetotail and a weak auroral zone may be detectable near the equator on the dark side of the planet. The topology of the magnetic field must have some peculiarities, notably the dayside funnel and the magnetotail with its plasmasheet cutting on the average across the ecliptic plane.

REFERENCES

(1) Gold, T. (1959). J. Geophys. Res. 64, 1219.
(2) Tsurutani, B.T. and Rodriguez, P. (1981). J. Geophys. Res., in press.
(3) Dungey, J.W. (1961). Phys. Rev. Lett. 6, 47.
(4) Levy, R.H., Petschek, H.E. and Siscoe, G.L. (1964). A.I.A.A. Jnl. 2, 2065.
(5) Reid, G.C. and Sauer, H.H. (1967). J. Geophys. Res. 72, 4383.
(6) Axford, W.I., Petschek, H.E. and Siscoe, G.L. (1965). J. Geophys. Res. 70, 1231.
(7) Axford, W.I. (1967). Space Sci. Rev. 7, 149.
(8) Akasofu, S.-I. (1968). Polar and Magnetic Substorms, D. Reidel Co., Dordrecht-Holland.
(9) Axford, W.I. (1969). Rev. Geophys. 7, 421.

(10) Schultz, M. and Lanzerotti, L.J. (1974). "Particle diffusion in the radiation belts", Springer, New York.

(11) Vasyliunas, V.M. (1972). in Earth's Magnetospheric Processes, ed. B.M. McCormac, D. Reidel, Dordrecht-Holland, 29.

(12) Swift, D.W. (1979). Rev. Geophys. and Space Sci. 17, 681.

(13) Sarris, E.T. and Axford, W.I. (1979). Nature 277, 460.

(14) Amano, K. and Tsuda, T. J. Geomag. Geoelec. 30, 27.

(15) Axford, W.I. (1972). NASA SP-308, 609.

(16) Axford, W.I. (1969). J. Geophys. Res. 73, 6855.

(17) Burch, J.L. (1979). Space Sci. Rev. 23, 449.

(18) Bridge, H.S. et al. (1979). Science 206, 972.

(19) McDonough, T.R. and Brice, N.M. (1973). Icarus 20, 136.

(20) Ip, W.-H. and Axford, W.I. (1980). Nature 283, 180.

(21) Goldreich, P. and Lynden-Bell, D. (1969). Astrophys. J. 156, 59.

(22) Hamilton, D.C., Gloeckler, G., Krimigis, S.M., Bostrom, C.O., Armstrong, T.P., Axford, W.I., Fan, C.Y., Lanzerotti, L.J. and Hunten, D.M. (1980). Geophys. Res. Lett. 7, 813.

(23) Axford, W.I. (1970). Particles and Fields in the Magnetosphere, ed. B.M. McCormac, D. Reidel, Dordrecht-Holland, 46.

(24) Axford, W.I. (1976). Proc. S.T.P. Symp. Boulder, 1, 270.

(25) Johnson, R.G. (1979). Rev. Geophys. and Space Phys. 17, 696.

(26) Anderson, K.A. and Lin, R.P. (1969). J. Geophys. Res. 74, 3953.

(27) Ness, N.F. (1978). Space Sci. Rev. 21, 527.

(28) Simpson, J.A., Eraker, J.H., Lamport, J.E. and Walpole, P.H. (1974). Science 185, 160.

(29) Suess, S.T. and Goldstein, B.W. (1979). J. Geophys. Res. 84, 3306.

(30) Sullivan, J.D. and Bagenal, F. (1979). Nature 280, 798.

(31) Smith, E.J., Davis, L. and Jones, D.E. (1976). Jupiter, ed. T. Gehrels, U. Arizona Press, 783.

(32) Krimigis, S.M., Carbary, J.F., Keath, E.P., Bostrom, C.O., Axford, W.I., Gloeckler, G., Lanzerotti, L.J. and Armstrong, T.P. (1981). J. Geophys. Res., in press.

(33) Ioanidis, G.A. and Brice, N.M. (1971). Icarus 14, 360.

(34) Mendis, A. and Axford, W.I. (1974). Ann. Rev. Earth and Planet. Sci. 2, 419.

(35) Hill, T.W. (1976). Planet. Space Sci. 24, 1151.

(36) Smith, E.J. et al. (1980). Science 207, 407.

(37) Judge, D.L., Wu, F.M. and Carlson, R.W. (1980). Science 207, 431.

(38) Krimigis, S.M. et al. (1981). Science, in press.

(39) Frank, L.A., Burch, B.G., Ackerson, K.L., Wolfe, J.H. and Mihalov, J.D. (1980). J. Geophys. Res. 85, 5695.

(40) Simpson, J.A. et al. (1980). Science 207, 411.

(41) Fillius, W., Ip, W.-H. and McIlwain, C.E. (1980). Science 207, 425.

(42) Van Allen, J.A., Thomsen, M.F., Randall, B.A., Rairden, R.L. and Grosskreutz, C.L. (1980). Science 207, 415.

(43) Brown, L.W. (1976). Astrophys. J. 207, L209.

(44) Armstrong, T.P., Axford, W.I., Bostrom, C.O., Fan, C.Y., Gloeckler, G., Krimigis, S.M., Lanzerotti, L.J. and Wilkens, D.J. (1975). Proposal for MJU-79 mission.

(45) Bridge, H.S. et al. (1975). Proposal for MJU-79 mission.

(46) Siscoe, G.L. (1971). Planet. Space Sci. 19, 483.
(47) Siscoe, G.L. (1975). Icarus 24, 311.
(48) Beard, D.B., unpublished.
(49) Axford, W.I. (1980). Proc. 10th Texas Symp. on Relativistic
          Astrophysics, Baltimore.

THE ROTATION OF URANUS

Richard M. Goody

Harvard University, Cambridge, MA 02138

HISTORICAL REVIEW

From the start of this century until the mid-1970's the rotation rate of Uranus was reported to be $10.8\ h$ in a retrograde sense (see, for example, Allen, 1955), but a cursory examination of the origin of this datum reveals that little confidence should be placed in it.

Three independent techniques for measuring the rotation rate are available, each very difficult and not including the most direct method of observing the motion of features across the disc. Visual observers have reported features from time to time (see Alexander, 1965, for a full historical account), but the mean diameter of Uranus is only $3.6\ arcsec$ and large high contrast features are rare in the visible spectrum, if they exist at all.

The three available methods are: use of theoretical interior models together with observations of the oblateness, f, and the gravitational moment, J_2; periodic fluctuations in the brightness; spectrographic measurements of Doppler shifts (line tilts). The first report of a rotation rate close to $11\ h$, based on a theoretical analysis of the planet's figure, was by Berstrand (1909; some of the early work is not easy to find and where it is only of historical importance I have relied upon Alexander's reports). At that time, there were no measurements of J_2 and no reliable data on the oblateness. The crucial early work was the spectrographic determination of $10.8 \pm 0.3\ h$ by Lowell and Slipher (1912). While these observers worked with exemplary care, a reanalysis of their data by Hayes and Belton (1977) shows no significant line tilts if all the data are taken together.

Soon after the work of Lowell and Slipher came the confirmatory work of Campbell from an analysis of brightness variations. A period of *10.82 h* was announced by E. C. Pickering in 1917, but Campbell did not publish until 1936 and, in the meanwhile, work in 1918 did not support his original findings. Finally, in the early era, came the spectrographic work of Moore and Menzel (1930) whose value of *10.84 ± 0.16* was quoted most widely until a year or two ago.

The spectrographic method is extremely difficult, as is demonstrated by a wide spread of results from recent observers using much more sophisticated equipment and analytical techniques than were available to Moore and Menzel. The brightness variations give a very precise result but sometimes of doubtful significance. It appears that mutual reinforcement may have occurred based upon confidence in the superior accuracy of an alternate technique. Whatever the reason for the agreement between Lowell and Slipher, Campbell, and Moore and Menzel, there is now no longer any doubt that the *11 h* period is incorrect.

The 1981 *Astronomical Almanac* now gives - *0.65d*? as the rotation period of Uranus. This is a period for which Robert Brown and I are responsible and it agrees with recent work on the figure of the planet. The query reflects editorial uncertainty because of some discordant measurements, to which I shall return.

One important feature of the rotation of Uranus was correctly established at an early date, namely, that the rotation axis lies close to the plane of the ecliptic with an inclination of *97°59'* (Fig. 1). Herschel noted in 1787 that the orbit of Oberon had a very high inclination. All of the rings and satellites of Uranus have similar inclinations and small precessions, facts which require that the orbits lie in the equatorial plane. This was first pointed out by Laplace in 1829, but was probably also appreciated by the Herschels. How the planet and its satellites reached this configuration is another matter (Greenberg, 1975), beyond the scope of this paper.

THE FIGURE OF URANUS

For an equilibrium rotating body the gravitational moment, J_2, and the oblateness, f, are to first order, related by (Cook, 1973):

$$J_2 = 2f/3 - m/3,$$

$$m = 4\pi^2 r^3/T^2 GM,$$

where T is the rotational period, r the radius of the planet, M the mass, and G the universal gravitational constant.

Given J_2 and f, therefore, the period can be calculated. J_2 can be determined from the precession of satellite orbits, but the precession is very small and the measurement was extremely difficult prior to the discovery of the rings. Oblateness must be measured directly. It is about 2%, and its magnitude was in doubt until recent years. Another relationship between J_2, f and m can be calculated given a model of the interior of the planet (Brown and Goody, 1980; Podolak, 1976; Podolak and Cameron, 1974). Houzeau, in 1856, used speculations about the interior structure of Uranus to obtain the first estimate of the period, between $7.25\ h$ and $12.5\ h$.

Figure 1. The appearance of Uranus and its satellites on 9 May 1981 (from the *Astronomical Almanac*, p. F62).

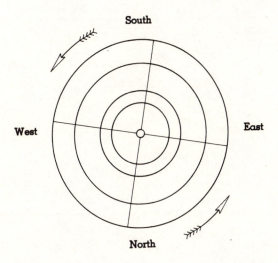

Despite a great deal of effort over the next 100 years, Houzeau's figure was not easily bettered. Figure 2 indicates the state of affairs in 1976. The two best measurements of J_2 from satellite orbit analysis are shown. The oblateness determination of Dollfuss (1970) is a summary of all ground-based work to date. That of Danielson, *et al.* (1972) is based upon measurements made by Stratoscope II. The full lines are for a range of interior models of Podolak (1976). On the basis of these data Podolak favored a period $\sim 18\ h$.

Occultation measurements on the rings of Uranus now allow very precise determinations of J_2. Nicholson, *et al.* (1978) find $J_2 = 3.43 \pm 0.02 \times 10^{-3}$, while more extensive work by Elliot, *et al.* (1980) gives $J_2 = 3.354 \pm 0.005 \times 10^{-3}$. Simultaneously, reliable data for the oblateness have become avilable. Franklin, *et al.* (1980) reanalyzed the Stratoscope II data and found $f = 0.022 \pm 0.001$. This result depends upon an understanding of the difference

Figure 2. The relationship between J_2, f and T for a range of interior models together with measurements available in 1976 (after Podolak, 1976).

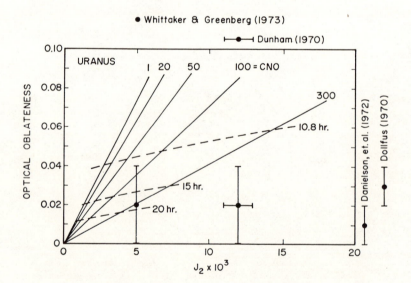

between polar and equatorial limb darkening, but the same criticism cannot be made of the work of Elliot, *et al.* (1981), which is based upon stellar occultations. Elliot, *et al.* find $f = 0.24 \pm 0.003$.

If we adopt Franklin, *et al.*'s value of f and Eliot, *et al.*'s value of J_2 we have
$$T = 16.7 \pm 0.5 \ h \ .$$

BRIGHTNESS VARIATIONS

The foregoing work on the figure of Uranus seems conclusive in its indications, but it is an indirect method and confirmation from a more direct approach is desirable. Brightness fluctuations offer one possibility. A periodic change in the brightness of a planet or satellite with a period on the order of tens of hours is most likely to be associated with the appearance and disappearance of bright or dark features on the limb. Our knowledge of the rotational period of Neptune is mainly based upon such measurements although for Neptune the variations are very large, up to *2 mag* in the J-K color index, according to Belton, Wallace and Howard (1981).

The history of brightness measurements on Uranus is summarized in Table 1. Most of the references can be found in Alexander (1965); the editorial comments are my own. The table speaks for itself and shows that no convincing case for brightness variations of Uranus has been established in the published literature.

The most accurate data are those of Lockwood and Thompson (1978), working at Lowell Observatory. They made measurements with filters at and adjacent to methane bands at *6190* and *7261 $\overset{\circ}{A}$*, because other measurements suggest that variability is at a maximum in strong methane bands. It is generally believed that clouds may occasionally form above the level of methane absorption and give rise to local brightening. Lockwood and Thompson conclude that there are no periodic variations in excess of *0.003 mag* except perhaps in the range of *23* to *25 h* where it is hard to reach a conclusion because of the difficulty of working with irregularly spaced data taken at 24-hour intervals.

Despite this difficulty, a period close to *24 h* is, in fact, claimed by a group working at the University of Texas. This work is reported in the abstract only (Smith and Slavsky, 1979): the methods are the same as those used by Slavsky and Smith (1978) for Neptune. I am indebted to David Slavsky for the following details.

Observations were made with filters in and adjacent to the 6190 $\overset{\circ}{A}$ methane band in Texas, Chile and South Africa. Periodic variations of the brightness of Uranus of *0.006 mag* were recorded in contrast to *0.002 mag* for a comparison star. The phase for simultaneous measurements in Chile and South Africa agreed in sidereal time but not in local time, and the entire data set is consistent with a period of *23.87 h.*

Table 1. The period of Uranus from brightness variations.

Date	Observer(s)	Period	Comments
1884-85	Muller	---	No variations
1915	Waterfield	*21 d*	Probably insignificant
1916-17	Campbell	*10.82 h*	*0.15 mag* variations
1918	Campbell	---	No variations
1921-26	Wirtz	---	No variations
1926	Perenago	*10.82 h*	Poor statistics
1926	Perenago	---	No variations
1927	Slavenas	*10.82 h*	Poor statistics
1928	Stebbins/Jacobsen	---	No variations
1928	Gussow	---	No variations
1934-35	Sterne/Calder	*10.82 h*	Marginal statistics
1976	Belton	*21.48 h*	Reanalysis of Campbell's data
1977	Lockwood/Thompson	---	Less than *0.003 mag* variations except for *23-25 h* period
1980	Smith/Slavsky	*23.87 h*	Unpublished; *0.006 mag* variations

SPECTROGRAPHIC METHODS

These methods are also more direct than those based upon the planetary figure. In their simplest form the spectrograph slit is placed along the equator and a measurement made of the tilt of a spectral line.

There are many pitfalls. Line tilts are typically a few degrees. For the observations of Moore and Menzel the diameter of Uranus at the photographic plate was $\sim$ *0.2 mm* and the exposure time was 1-2 hours. The lateral distance between the two ends of the line averaged *5 μm* or only 0.5 grains, depending upon the characteristics of the film used.

Modern instruments give great improvements. In 1976 and 1977, Robert Brown and I (Brown and Goody, 1977; 1980) worked with the KPNO 4-meter telescope with a Cassegrain echelle to obtain both high spectral dispersion and a large image. The detector was a Kron camera having extremely fine grain combined with linearity over a wide dynamic range. Most importantly, we took steps to allow for the effect of seeing on the recorded line tilts.

The correction function is shown in Fig. 3. The recorded line tilt decreases as the seeing disc increases in size. For *s arcsec* the recorded line tilt must be increased in the ratio $G(0)/G(s)$ before interpretation as a Doppler shift. For a seeing disc of *2 arcsec* this ratio is about 1.6. For the long exposures of Moore and Menzel the correction factor (which they did not apply) must be at least this great, although this only serves to increase the discrepancy between their value and all modern determinations.

Seeing can be measured from scans across the spectrum in continuum regions, but use of this result requires that seeing and guiding errors be isotropic. To eliminate the typical bias between errors in RA and Dec we employed automatic guidance on all occasions.

In 1976 we measured the tilts of 23 lines and obtained a rotational period of *15.57 $\pm$ 0.80 h*, while in 1977 we measured more than 600 lines for a period of *16.26 $\pm$ 0.34 h*. Both data sets are consistent with

$$T = 16.16 \pm 0.33 \ h \ .$$

We were also able to confirm that the rotation axis corresponded with the pole of the satellite orbits.

The work was only one of a number of simultaneous attempts to improve spectrographic determinations of Uranus' period. Trauger, Roessler and Münch (1978) used a whole-disc approach which avoids seeing corrections. It requires a knowledge of incoherent scattering processes, however, about which we cannot be confident. Trafton (1977) employed an ingenious variant of the normal spectrographic technique which is, however, subject to large errors. I shall therefore restrict my discussion to the work of Münch and Hippelein (1980) and Hayes and Belton (1977).

Figure 3. Seeing correction to line tilt based on measurements of synthetic spectra. For perfect seeing $G(0)$ is 7.65×10^{-2} (after Brown and Goody, 1977).

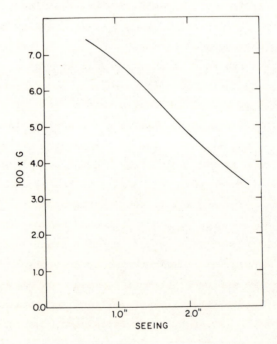

The former obtain a period of $15.0\,^{+4.0}_{-2.6}$ h, consistent with our work, and they provided a quasi-analytical basis for the seeing correction. Hayes and Belton obtained 24 ± 3 h, consistent with the work of Smith and Slavsky, but irreconcilable with our determination.

Hayes and Belton used essentially the same equipment as we did and employed a similar method of analysis. It is difficult to avoid the conclusion that one piece of work or the other contains numerical errors. With elaborate numerical algorithms such errors are regrettably easy to make and hard to detect. In a subsequent paper, Belton, Wallace, Hayes and Price (1980) mention "two serious sources of error" in the earlier work but state that their results for Uranus are unaffected.

THE ROTATION OF URANUS

The weighted mean of our data and that obtained from the planetary figure are

$T = 16.31 \pm 0.27$ h,

The values of J_2 and f are now firmly established. For the rotational period to be wrong the equilibrium theory of the figure must be inapplicable. It works very well for all other planets for which data exist, except for the Moon.

The work of Slavsky and Smith cannot be evaluated until it is published and the only established discrepancy with the above result is, therefore, the work of Hayes and Belton. Unfortunately, it will be a long time before the spectrographic work can be repeated successfully, even assuming that investigators will exist with the desire to do so. The aspect of the planet is becoming increasingly unfavorable. In 1985 the North Pole will point directly toward the Sun and seven or eight more years must elapse before there is a substantial component of the rotation vector orthogonal to the line of sight.

The possibility of detecting the motion of features across the disc remains. Nisenson, *et al.* (1981) have reported on the use of speckle imaging techniques to obtain images of Titan with a resolution of 0.29 $arcsec$. With this equipment Uranus can be imaged in the *6190 Å* methane band and with more suitable image intensifiers, also in the *7261 Å* band. If large cloud systems appear above the level of methane absorption it may be possible to detect them.

Finally, the Voyager II flyby in January 1986 may show features on the disc, in which case the controversy over the Uranus rotation period should be finally resolved.

ACKNOWLEDGEMENTS

I am indebted to Robert Brown for past collaborations. My work is supported by the National Aeronautics and Space Administration under Grant NGL-22-007-228.

REFERENCES

Alexander, A. F. O'D (1965). *The Planet Uranus: A History of Observation, Theory and Discovery*. Faber and Faber, London.

Allen, C. M. (1955). *Astrophysical Quantities*. Athlone Press, London, p. 157.

Belton, M. J. S., Wallace, L. and Howard, S. (1981). The periods of Neptune: Evidence for atmospheric motions. Submitted to *Icarus*.

Belton, M. J. S., Wallace, L., Hayes, S. H. and Price, M. J. (1980). Neptune's rotation period: A correction and a speculation on the difference between photometric and spectrographic results. *Icarus* 42, 71-88.

Brown, R. A. and Goody, R. M. (1977). The rotation of Uranus. *Ap. J.* 217, 680-687.

Brown, R. A. and Goody, R. M. (1980). The rotation of Uranus II. *Ap. J.* 235, 1066-1070.

Campbell, L. (1936). The rotation of Uranus. *Harvard C. O. Bull.* 904, 32-35.

Cook, A. H. (1973). *Physics of the Earth and Planets*. Wiley, New York, pp. 35, 132.

Danielson, R. E., Tomasko, M. G. and Savage, B. D. (1972). High resolution imagery of Uranus obtained by Stratoscope II. *Ap. J.* 178, 887-900.

Dollfuss, A. (1970). Diamètres des planètes et satellites. In *Surfaces and Interiors of Planets and Satellites* (A. Dollfus, Ed.). Academic Press, New York, pp. 46-139.

Elliot, J. L., French, R. G., Foregl, J. A., Elias, J. H., Mink, D. and Liller, W. (1981). Orbits of nine Uranian rings. Submitted to *Ap. J.*

Franklin, F. A., Avis, C. C., Columbo, G. and Shapiro, I. I. (1980). The geometric oblateness of Uranus. *Ap. J.* 236, 1031-1034.

Greenberg, R. (1975). The dynamics of Uranus' satellites. *Icarus* 24, 325-332.

Hayes, S. H. and Belton, M. J. S. (1977). The rotational periods of Uranus and Neptune. *Icarus* 32, 383-401.

Lockwood, G. W. and Thompson, D. T. (1978). A photometric test of rotational periods for Uranus and time variations of methane-band strengths. *Ap. J.* <u>221</u>, 689-693.

Lowell, P. and Slipher, V. M. (1912). Spectroscopic discovery of the rotation period of Uranus. *Lowell Obs. Bull.* <u>2</u>, No. 3.

Moore, J. H. and Menzel, D. H. (1930). The rotation of Uranus. *Pub. Astron. Soc. Pac.* <u>42</u>, 330-335.

Münch, G. and Hippelein, H. (1980). The effects of seeing on the reflected spectrum of Uranus and Neptune. *Astron. Astrophys.* <u>81</u>, 189-197.

Nicholson, P. D., Pensson, S. E., Matthews, K., Goldreich, P. and Neugebauer, G. (1978). The rings of Uranus: Results of the 1978 April 10 occultation. *Astron. J.* <u>83</u>, 1240-1248.

Nisenson, P., Apt. J., Goody, R. and Horowitz, P. (1981). Radius and limb darkening of Titan from speckle imaging. Submitted to *Astron. J.*

Podolak, M. (1976). Methane rich models of Uranus. *Icarus* <u>27</u>, 473-476.

Podolak, M. and Cameron, A. G. W. (1974). Models of the giant planets. *Icarus* 22, 123-148.

Slavsky, D. and Smith, H. J. (1978). The rotation period of Neptune. *Ap. J.* <u>226</u>, L49-52.

Smith, H. J. and Slavsky, D. B. (1979). Rotation period of Uranus. *B.A.A.S.* <u>11</u>, 568.

Trafton, L. (1977). Uranus' rotational period. *Icarus* <u>32</u>, 402-412.

Trauger, J. T., Roessler, F. L. and Münch, G. (1978). A redetermination of the Uranus rotation period. *Ap. J.* <u>219</u>, 1079-1083.

AN INTRODUCTORY REVIEW OF OUR PRESENT UNDERSTANDING OF THE STRUCTURE AND COMPOSITION OF URANUS' ATMOSPHERE

Michael J. S. Belton
Kitt Peak National Observatory †
Tucson, Arizona 85726

Measurements of the gross properties of Uranus are enough to show at it is a planet with quite unique characteristics. The best estimates its radius (24900 km; Danielson, Tomasko and Savage, 1972) and of its ss (14.54 Earth masses; Klepczynski, Seidelmann and Duncombe, 1970) imply interior structure and overall composition that set it clearly apart om Jupiter and Saturn (Hubbard and MacFarlane, 1980).

The best estimates of the maximum outflow of internally generated at (< 250 ergs cm^{-2} sec^{-1}; cf. Wallace, 1980) and its peculiar axial ientation (98° obliquity; Dunham, 1971) set it apart from its otherwise)erficially similar sister planet, Neptune, when considering problems of 1ospheric structure.

In this review, I will attempt to summarize the enormous progress at I believe has been made during the past decade (the average number of search papers regarding Uranus' atmosphere has increased from an average one per year in the late 1960's to an average of ∿ 13 papers per year 1980) in understanding the planet's spectrum in terms of the probable emical and physical nature of its atmosphere.

Essentially, all the work which has been done so far is *exploratory* nature and I cannot report that a concensus on the mean structure and 1position of the planet's atmosphere has yet been achieved; there are, :eover, several important areas that need to be further explored before :h a concensus can be expected. I am thinking particularly of questions icerning the time dependence of atmospheric structure which might occur a result of the extreme seasonal characteristics of Uranus. On Uranus, Like other planets, the polar regions receive an average of π/2 more solation than the equatorial regions. It is, nevertheless, possible to entify the general domain of the mean structure and composition and, rhaps more significantly, identify some of the critical observations on .ch future advances may depend.

)erated by the Association of Universities for Research in Astronomy, ic., under contract with the National Science Foundation.

   I begin by presenting a thumbnail sketch of our current view of
atmosphere in the context of a comparison with Neptune's atmosphere which
in the past, has often been assumed to have similar properties.   This
similarity of the two planets is now, I believe, very much in question.

1.   A Thumbnail Sketch

   So far as we know, the primary atmospheric constituents of the
visible atmosphere (H_2 and He) represent only 11 percent of the planet's
mass (Hubbard and MacFarlane, 1980) and, *if* an interface between this atm
phere and the ice-rock mixture that makes up the bulk of the planet, in f
exists, is about 7800 km deep with a base pressure near 221,000 bars at a
temperature of 2500°K.

   Of the sensible atmosphere itself, our observations have probed,
albeit crudely in many cases, the region from $\sim$ 10 μ bars to near the 100
bar pressure level.   Above 10 μ bars, we are in the realm of aeronomical
conjecture which I shall not review although some exploratory work has be
published by McElroy (1973) and by Capone et al (1977).   100 bars is abou
the deepest level that it has been possible to probe with microwave radic
metry (21 cm observations; Briggs, 1973).

   The Uranian stratosphere (10 μ bar - 0.05 bar) is surprisingly
cold ($\sim$ 95°K) while that of Neptune has a temperature ($\sim$140°K) and emissi
properties that are more in concert with our experience at Jupiter and
Saturn.   This may imply small but significantly different tropopausal str
tures on the two planets, in spite of their almost identical effective
temperatures, since it is the tropopause temperature which probably contr
the stratosphere methane content and, thereby, its thermal balance with
solar insolation.   The meaning of "significant" can be inferred by noting
that only a 3°K change in the assumed effective temperature could lead to
a factor of two in the inferred stratospheric temperature (Wallace, 1975)
on account of the sensitivity of the vapor pressure of CH_4 on temperature
Evidence that such differences do exist in the tropopausal region ($\sim$ 50 m
bar level) is evidenced by the presence of cloud and "weather" at this
level on Neptune (Joyce, Pilcher, Cruikshank and Morrison, 1977) but not
on Uranus.

   Between 0.1 and 10.0 bar, where the bulk atmosphere is in most
effective radiative contact with space, the controlling thermal opacity
(collisionally induced translational transitions in H_2 - He) and the "eff
tive" temperature are essentially identical on the two planets.   As a con
sequence, the structures in this region should be very similar; a fact to
which the visible and near infrared spectrum atests.   In this region, the
scale height is about 40 km and the lapse is near 0.8°K km^{-1}; methane con
denses in this region, solar heating has its peak value, and there are

ndications of a system of layered clouds.

At pressures greater than a few tens of bars ($\sim$ 30 bar; cf. Wallace, 1980) the thermal structure is very sensitive to the actual cloud structures that are present (they control the deposition of solar heat) and the internal heat flux. Since it seems that Neptune's internal heat flow ($\sim$ 300 ergs cm^{-2} sec^{-1}) is at least two times that of Uranus (and perhaps much more) then it seems to me quite possible that the two planets might exhibit different structures at these levels. On the other hand, the microwave spectra (1 - 21 cm) which are thought to probe these regions on both planets are very similar (cf., Morrison and Cruikshank, 1973) and, even though the shape of either spectrum has yet to have a satisfactory explanation (Gulkis, Janssen, and Olsen, 1978), this probably means similar atmosphere structures at these depths (10 - 100 bar).

Before moving into a more detailed discussion and since this paper is in the nature of an introductory review, I would now briefly like to note a few of the speculations that have crept into the research literature and which are now known to be definitely incorrect. This should be helpful to new readers of the literature and also helps to salve my own conscience, since I helped perpetuate some of them. The idea that extended wings in collisionally induced dipole transitions in H_2 provide substantial opacity throughout the near infrared (Belton, McElroy, and Price, 1971) is no longer necessary or useful; the idea that simultaneous transitions in a CH_4 - H_2 mixture have a role in producing the near infrared spectrum (Danielson, 1974) has been disapproved by new and extensive laboratory spectroscopy by Fink, Benner and Dick (1977) and by Giver (1978); the idea that no detectable aerosols exist in the Uranus atmosphere (Belton et al, 1971) has been disproved by proper treatment of radiative transport in the H_2 quadrupole lines and observations of the (3 - 0) S(0) collisionally induced feature of H_2 at 8250Å (Trafton, 1976; Belton and Spinrad, 1973); confusion in the literature on the value of the transition movements of lines in the (4 - 0) band of H_2 (cf., Wallace, 1980) has now been cleared up with further laboratory work by Brault and Smith (1980) and the measured values are close to the theoretically computed value (Dalgarno and Allison, 1969). Thus, Trafton's (1976) work in interpreting the H_2 spectrum of Uranus remains valid; finally, speculations that have surfaced from time to time regarding the presence, in large quantities, of a spectroscopically inert gas other than He (eg., from the studies of sharp H_4 lines near 6819Å) all appear now to be without any real basis.

A Stroll through Uranus' Spectrum

As an introduction to a discussion on structure, let me now briefly review the character of the Uranian spectrum from whence the major part of our knowledge comes. This emphasis, of course, is not meant to

deny the exceedingly important contributions from photometry (albedo;
Younkin, 1970); from occultation measurements (stratospheric temperatures
Dunham, Elliot and Gierasch, 1980); from polarimetry (eg., Michalsky and
Stokes, 1977; presence of upper atmospheric aerosols); from imagery
(dimensions; limb brightening/darkening; Danielson et al, 1972; Sinton,
1972); and from center-to-limb observations (atmospheric inhomogeniety;
CH_4 saturation; Price and Franz, 1979; Pilcher et al., 1979).

In the ultraviolet, the shortest wavelengths at which, to my
knowledge, the planet has been observed is 1800Å by the ANS satellite
(Savage, Cochran and Wesselius, 1980). The planetary spectrum between
1800Å - 4200Å, which is superposed on the reflected solar spectrum, is
dominated by the effects of rotational Raman transitions in H_2 (Belton,
Wallace and Price, 1973) and a diffuse absorption continuum whose origin
is, at present, undetermined. At one time, I considered that observation
of the Raman effect would impose significant boundary conditions on the
distribution and amount of aerosols in the visible atmosphere, but this h
not turned out to be the case and almost any of the many models considere
so far for the Uranus atmosphere will satisfy the observations. The Rama
transitions do, however, make a significant statement about the ortho/par
ratio of H_2 in the atmosphere.

The underlying absorption may well be the same "Axel" dust that
exists in the upper atmospheres of Jupiter and Saturn - a photochemical
detrius - but this is ad hoc and some of the absorption may be associated
with one of the atmospheric cloud layers.

From 4200Å out to nearly 5μ, the spectrum is dominated by a complex progression of CH_4 bands. Recent and extensive laboratory work by
Lutz, Owen, and Cess (1976), by Fink, Benner, and Dick (1977), by Giver
(1978), and by Silvaggio (1977); and some measurements at low temperature
(in the stronger bands) have enabled the first quantitative attempts of
explaining the planetary spectrum in terms of the complex coupling betwee
the abundance of CH_4 and atmospheric structure (Wallace, 1980). The so-
called "blue" bands near 4800Å are found to place the most significant
constraints on the deep CH_4/H_2 mixing ratio since they are weak enough
for the radiation field to probe below the level of CH_4 saturation; the
stronger bands near 1μ put strong constraints on the abundance of upper
atmospheric aerosols and the He/H_2 ratio but not on the abundance of CH_4;
finally, but in a less constructive way, the weak interband absorption in
CH_4 severely limits the interpretation of the (3 - 0) collisionally induc
dipole absorption of H_2 near 8250Å which is very important for constraini
cloud structure (Trafton, 1976) and measurements in the laboratory at lo
temperature are urgently needed.

Superposed in the CH_4 spectrum in this region is the relatively sparce spectrum of H_2 and HD. HD <u>has</u>, I think, been detected (Trafton and Ramsey, 1980; Trauger et al, 1977) in the (5 - 0) band and the mixing ratio with H_2 is close to $4 \cdot 8 \times 10^{-5}$, similar to Jupiter and Saturn. This result evidently must be taken into consideration in condensation scenarios for the planet (Hubbard and MacFarlane, 1981) for enhancements of a factor of two or more in Uranus and Neptune are expected. Perhaps on Uranus, D is still "trapped" in a partially solid ice ball? Observations of HD in Neptune are obviously of great interest.

The observed H_2 spectrum consists of six lines of the quadrupole spectrum (5 - 0, 4 - 0, and 3 - 0 bands) and one line of the pressure induced dipole spectrum (S(0) in the 3 - 0 band). The relative strengths of lines from odd and even rotational levels indicate that the H_2 rotational populations are close to thermal equilibrium for the temperatures in Uranus' upper atmosphere. (This is also indicated by the Raman spectrum.)

The H_2 spectrum has turned out to be the crucial element in understanding cloud structure in the visible part of the atmosphere (cf., Trafton, 1976). The (3 - 0) quadrupole and dipole lines require a base cloud at a depth of $\sim$ 500 km - amagats of H_2 ($\sim$ 5 bar level); while the (4 - 0) lines require, in addition, a thin, tenuous, haze high in the atmosphere ($\tau \sim 0.3$ at 6400Å). The haze is ineffective if placed low in the atmosphere while its highest possible altitude is restricted by the observed albedo in the bottom of strong CH_4 bands. The 1 bar level appears to be appropriate for this haze. Uncertainties in the observed strength of the (4 - 0) lines (Smith, Macy, and Pilcher, 1980) lead to considerable uncertainties in the properties of this elevated haze layer, although Wallace (1980) finds it an essential component in balancing the predicted strength of the CH_4 "blue" band spectrum with the (3 - 0) dipole feature.

It might be expected that at 5μ, both Uranus and Neptune should easily be detected in reflected sunlight for molecular and aerosol scattering represents the only known and unexpected opacity. The planet has been detected (Gillett and Rieke, 1977; Macy, Sinton, and Beichman, 1980), but with the very low albedo of $\sim$ 0.01. Macy et al suspect CH_3D and PH_3 as being the source of this absorption and spectra in this region are urgently needed.

Beyond 5μ is the undisputed domain of thermal radiation and the observed spectrum is dominated by collisionally induced rotational and translational transitions in H_2. Stratospheric emissions of stratospheric hydrocarbons at 7.8 and 12 μ have not yet been observed on Uranus and new efforts are clearly needed since I am not aware of any work since that of Gillett and Rieke (1977) or Macy and Sinton (1977). The stratosphere is

not a total unknown as a result of successful observations of a grazing
occultation of SAO 158687 by the planet from the Kuiper Airborne Observatory in 1977 (Dunham, Eliott and Gierasch, 1980) for a reasonable mixture of H_2 and these observations indicate a temperature of $\sim$ 95° K at
the 10μ bar level considerably cooler than is the case for the other giant
planets.

Between 10 and 30μ where opacity due to the S(0) and S(1) lines
in H_2 (collisionally induced dipole rotational transitions) should allow
probing of the upper troposphere and possibly the lower stratosphere,
the observations are broad band radiometric measurements and are too crude
to establish whether an inversion definitely exists in the 0.01 bar
region (cf., Courtin, Goutier, and Lacombe, 1978), but they suffice to
demonstrate the reasonableness of radiative-convective model atmospheres
(Wallace, 1980).

From 30μ to a few mm, the spectrum provides what is perhaps its
most important constraint on atmospheric structure. This is because the
bulk of the planetary thermal radiation emerges from the atmosphere in
this spectral region where the opacity is controlled by collisionally
induced dipole absorption between translational states of H_2 and He.
Deep in the atmosphere, CH_4 also plays an important role in the magnitude
of this opacity (Wallace, 1980) but in the region of the visible atmosphere, the abundance of CH_4 is depleted by condensation and is, therefore,
not a factor in determining structure (Trafton, 1972). The various
measurements show that the effective temperature is within 2 degrees of
58°K (Courtin et al, 1978; Stier et al., 1978) which restricts the internal outflow of internal heat to less than 250 ergs cm^{-2} sec^{-1}. Theoretical estimates of the internal heat flow range from 100 erg cm^{-2} sec^{-1}
to 10 ergs cm^{-2} sec^{-1} (Danielson, 1975) and could possibly be lower.

H_2 - He opacity controls the spectrum to about 1 cm wavelength
at which point the spectrum becomes exceedingly interesting and problem-
atical. In the region between 1 - 21 cm, the spectrum was anticipated
to show brightness temperatures near 150°K and a shape that reflected the
powerful absorption associated with inversion transitions in NH_3. A
"flat" spectrum was expected because of the high pressure ($\sim 10^3$ bar) at
which the spectrum would be formed and as a result of the vapor pressure
characteristics of NH_3 which cause a rapid increase in opacity over a
limited vertical extent in the atmosphere. It was, therefore, a great
surprise (albeit delayed) when Gulkis (1975) pointed out the fact that the
spectrum is unexpectedly hot at 2 - 10 cm; a surprise that was compounded
when Klein and Turanago (1978) found the micro wave spectrum to be (at
least at the present time) changing with time. Even at the present time
the shape of the Uranus microwave spectrum from 2 - 21 cm, which is
shared by Neptune, has defied adequate explanation.

3. Structure and Composition

a. General Considerations

In a deep, cold, atmosphere like that which exists on Uranus, both the scattering of radiation and compositional inhomogeniety are of first order importance in the process of spectral line and band formation. As a result, we have learned by experience that the compositional problem and the structural problem are inextricably mixed. Most of the work to date has been of an exploratory nature; attempts to understand in the widest sense the complex interactions between assumed composition, cloud structure, and thermal structure. With the work of Danielson and his associates (1975, 1977), Trafton (1976), Courtin, Gautier and Lacombe (1978), and Wallace (1980), we have, I believe, arrived in the correct domain of structure and composition but, as yet, still lack a firm quantitative understanding.

Schematically, there are two approaches available for probing the atmospheric structure: "inverting" the observed spectral data on the basis of a set of rather restrictive and ad hoc assumptions (composition, behavior of the structure near upper and lower boundaries of the region probed) and the equation of radiative transfer; and, secondly, construc- ting sets of highly parameterized but physically consistent and reasonable radiative-convective models and comparing the predicted spectrum with observations. The inversion method has been applied by Courtin et al (1978) and serves to give a crude idea of the general domain of possible models. The inversion method is, I believe, unsuitable for more refined work at the present time because of the very poor quality of the 10 - 300 μ data that must be used (but see G. Orton's contribution to this collo- quium).

The "modelling" method is technically more stable and forgiving and, in my view, is much more informative as it clearly exposes the influ- ences of the physical processes that are at work and also help to identify whatever shortcomings exist in the observational data. This is the tech- nique used by Danielson and his associates (1975, 1977), by Trafton (1976), by Macy (1979), and most recently by Wallace (1980). The remain- der of this review is concerned with this latter approach.

In creating a model, the following steps are normally employed:

(A) Assumptions regarding composition and internal heat flow

(B) First guess at a thermal structure (i.e., the dependence of temperature on pressure throughout the atmosphere)

(C)   Imposition of cloud characteristics

(D)   Calculation of solar heat deposition

(E)   Correction of initial thermal structure by iteration on
      step (B) to achieve a constant net energy flux condition
      throughout the atmosphere.

In step (A), the molecules and atoms we expect to be important
for the visible atmosphere are a subset of those seen on Jupiter, but
primarily H_2, He, CH_4 and NH_3. Prinn and Lewis (1973) have noted that
H_2S may be an important molecule in the lower regions of the visible
atmosphere and that it could be substantially enhanced (relative to NH_3)
over solar proportions as a result of some possible scenarios for the
planet's origin. An He/H_2 ratio near 0.18 (e.g. Wallace, 1980) is used in
the modern models, and only the (very significant) effects of CH_4 condensation need be considered in structural models of the visible atmosphere.
The CH_4/H_2 mixing ratio is taken as a free parameter in most models and
the most satisfactory models have it enhanced some 40 times over the
solar value ($\sim 7 \times 10^{-4}$). NH_3 and H_2S may have significant effects on
the microwave spectrum but are apparently unimportant in structure calculations. Hunten (1981) has considered the effects of H_2O on the
structure but these are all confined to depths below 100 bars which are
currently outside the limits of the observable atmosphere.

Limits exist for line of sight abundances of NH_3 and H_2S (Fink
and Larson, 1979) but a proper interpretation of these limits in terms
of deep mixing ratios has not been attempted and would be, I think, an
exceedingly difficult task.

In step (C), two approaches have been used. In the first, a
minimum of clouds and hazes are placed in the model atmosphere where need
in order to get a satisfactory rendition of the observed spectrum. Walla
(1980) who employs variations on Trafton's (1976) cloud models uses this
approach as does Macy (1979). When a satisfactory physical structure is
achieved, the models are then examined to see if the assumed cloud layers
can be identified with potential condensates in the atmosphere. In these
models, CH_4 forms a high haze near 1 bar with a thick cloud base of H_2S
or NH_3 near the 5 bar level. In the second method, clouds are presumed
a priori to occur where an assumed chemical component is predicted to condense. The models of Danielson (1975) are of this type and are essential
single CH_4 cloud models.

In step (D), great advances have been made by Macy (1979) and Wallace (1980) through the use of laboratory spectra of CH_4 (Fink et al, 1977); Giver, 1978) in computing solar heat deposition as a function of depth in the atmosphere. The heating (in terms of ergs/gm/sec) peaks just above the 1 bar level, but its influence on structure may extend well below the 10 bar level depending on the precise nature of the lower cloud (cf., Wallace, 1980) and the magnitude of the internal heat flow.

Danielson (1975) was the first to explore the influence of heat deposition on structure with the help of a highly parameterized heating function. He was able to show that deep radiative structures might be possible for Uranus but I believe that models are now mainly of historical interest for they are all too cold and fail to give a reasonable account of the microwave spectrum between 0.1 and 1 cm.

Step (E), which is the heart of the modelling method, involves the calculation of thermal opacity. Now usually based on the work of Birnbaum, (1978) and numerical tests of the local buoyant stability of the atmosphere. The latter are dependent on the assumed composition through the specific heat of the atmospheric mixture. Both the assumed proportion of CH_4 and the ability of H_2 to convert between its ortho and para varieties is important in the calculation of specific heat. At the low temperatures in the Uranus visible atmosphere so called "equilibrium" H_2 (in which transitions between ortho and para states are assumed unhindered) and "normal" H_2 have very different specific heats, and the kind of H_2 which is assumed leads to quite different stability conditions. Wallace (1980) has made a major point of including this problem in his calculations and carries three assumptions as to the state of H_2: (i) "normal" H_2 in which the ortho-para ratio is effectively assumed to be that imposed at high temperature deep in the atmosphere, (iii) "equilibrium" H_2 in which ortho-para conversion is unhindered by quantum mechanical restriction at any temperature, and (iii) "intermediate" H_2, first postulated by Trafton (1967), in which the ortho-para ratio responds (presumably on a very long time scale) to conversion at the mean local temperature but does not respond on the short time scales of local convective processes in the atmosphere. Thus, "intermediate" hydrogen reflects a population of rotational states similar to "equilibrium" H_2 but has a specific heat similar to "normal" hydrogen. I think that in this application to Uranus Wallace has convincingly shown that it is to this "intermediate" case that H_2 prefers to conform in the atmosphere of the outer planets.

(b) Wallace's (1980) Models

The essential ingredients of the successful model, chosen to give the widest satisfaction to the complete set of observations, rather than precise agreement to any particular observation, are as follows:

(i)   a Trafton (1976) type cloud and haze structure consisting
      of a diffuse, highly reflective, but optically thin haze of
      probably CH_4, near 1 bar which is separated from a thick
      (in visual wavelengths) cloud base near the 5 bar level.
      This lower cloud takes on an albedo to sunlight of 0.98
      (and is probably H_2S or NH_3).

(ii) a ubiquitous, lightly absorbing haze (photochemical
      "detrius"?)

(iii) "intermediate" hydrogen

(iv) a mixing ratio for methane of 0.03 relative to H_2.

The best models (shown in Fig. 17 of Wallace, (1980) represent,
in my view, the best approximation to the mean atmospheric structure on
Uranus between the 0.1 and 30 bar level available at this time. Below
30 bars, the structure is uncertain and depends partly on what is assumed
about the magnitude of the internal heat flux and partly on the optical
properties of the base cloud at 5 bars. Above 0.1 bars, the structure
is similarly uncertain depending on the precise nature and mechanisms
of any tropopausal "cold trap", on the particular nature of the photo-
chemistry, and on whatever energy transport and disipation mechanisms
exist in that region.

Wallace's comparison of his models with observational data is
instructive: The models predict Bond albedos and effective temperature
and since all of them give an adequate account of these quantities
(Bond albedo between 0.315 and 0.361; effective temperatures between
56.9 and 60°K) it turns out that these parameters do not help
discriminate between the models but merely represent an overall
check of their validity.

The predicted IR spectrum between 10 and 300μ turns out to be
a poor discriminator between models, primarily because of the very poor
quality and the inconsistency in the data set.

It is the microwave spectrum between 1 - 10 mm and in the visual
and near infrared spectrum that the most leverage to discriminate between
the various models is found. The mm observations rule out all "equilibrium" H_2 models and those "intermediate" H_2 models with too much CH_4.
The CH_4/H_2 ratio is restricted to less than 0.1.

In the visible and near infrared reflection spectrum, the CH_4 "blue" bands put a lower limit on the amount of methane and require $CH_4/H_2 > 0.01$; while a combination of the H_2 spectrum and the strong CH_4 bands near $.8\mu$ set the character of the cloud structure as explained previously.

The models place little constraint on the internal heat flux although the comparison with the microwave spectrum seems to prefer the models with higher internal heat flow.

However, this is barely a significant result, if at all.

4. Unsolved Problems, Speculation, and Future Observations

The outstanding contemporary problems in structure and composition that I believe need to be addressed in the immediate future are as follows:

(i)   Why does Neptune have so much more cloud activity in the upper troposphere then Uranus?

(ii) What seasonal variations occur in the Uranus atmosphere; particularly near the tropopausal minimum?

(iii) What is the explanation of the shape and variability of Uranus' microwave spectrum between 1 and 21 cm?

(iv) What is the explanation of the apparently low atmospheric deuterium and nitrogen abundances?

The first three problems are probably related and will require high quality observations in the 5 - 30 μ region over an extended time base together with more occultations and observations particularly from Voyager. Problems (iii) and (iv) may also be interrelated. The microwave spectrum at wavelengths longer than 2 cm seems to demand either a local or overall depletion of NH_3 by a factor of 100 relative to solar abundance depending on the particular hypothesis that is made regarding the deep atmospheric structure. Excluding what seems to be clearly unacceptable hypotheses involving ionospheric emission and magnetospheric emission, the following ideas have been proposed as possible explanations of the shape of the microwave spectrum:

(a)   depletion of NH_3 coupled with a deep, roughly isothermal region below 30 bars ($\sim 250°K$)

(b)   a "surface" near the 280°K level (plus a depletion of NH_3).

(c)   chemical removal of NH_3 at the 150°K level by a super abundance of H_2S.

Idea (a) seems unacceptable because it does not provide a basis for an explanation of Neptune's microwave spectrum that has similar properties (Morrison and Cruikshank, 1973). Idea (b) might be acceptable if the equation of state of cosmic ices can accommodate an ice ball as large as Uranus and still meet the constraints of the mean density, as well as accommodate Neptune's large internal heat flow. In the context of this idea, the low N and D abundances would be explained by keeping these atoms largely trapped within the "ice ball." Idea (c) may work for both planets but leaves the deuterium problem unresolved. This idea also presents an untested hypothesis regarding the sulphur content of the atmosphere. Observations of HD on Neptune are needed as are more sensitive attempts to detect H_2S.

REFERENCES

Belton, M. J. S., McElroy, M. B., and Price, M. J. (1971)
    "The Atmosphere of Uranus," Astrophys. Journ., Vol. 164,
    191 - 209.

Belton, M. J. S., and Spinrad, H. (1973) "H_2 Pressure-Induced
    Lines in the Spectra of the Major Planets," Astrophy. Journ.,
    Vol. 185, 363 - 372.

Belton, M. J. S., Wallace, L., and Price, M. J. (1973) "Observation
    of the Raman Effect in the Spectrum of Uranus," Astrophys.
    Journ. (Letters), Vol. 184, L143 - L146.

Birnbaum, G. (1978) "Far Infrared Absorption in H_2 and H_2 - He
    Mixtures," J. Quant. Spectros. Radiat. Transfer, Vol. 19,
    51 - 62.

Brault, J. W., and Smith, W. H. (1980) "Determination of the H_2
    4 - 0 S(1) Quadrupole Line Strength and Pressure Shift,"
    Astrophys. Journ. (Letters), Vol. 235, L177 - L178.

Briggs, F. H. (1973) "Observations of Uranus and Saturn by a New
    Method of Radio Interferometry of Faint Moving Sources,"
    Astrophys. Journ., Vol. 182, 999 - 1011.

Capone, L. A., Whitten, R. C., Prasad, S. S., and Dubach, J.
    (1977) "The Ionosphere of Saturn, Uranus, and Neptune,"
    Astrophys. Journ., Vol. 215, 977 - 983.

Courtin, R., Gautier, D., and Lacombe, A. (1978) "On the Thermal
    Structure of Uranus from Infrared Measurements," Astron.
    and Astrophys., Vol. 63, 97 - 101.

Dalgarno, A., and Allison, A. C. (1969) "Rotation-Vibration
    Quadrupole Matrix Elements and Quadrupole Absorption
    Coefficients of the Ground Electronic States of H_2, HD and
    D_2," J. Atmos. Sci., Vol. 26, 946 - 951.

Danielson, R. E., Tomasko, M. G., and Savage, B. D. (1972)
    "High Resolution Imagery of Uranus Obtained by Stratoscope
    II," Astrophys. Journ., Vol. 178, 887 - 900.

Danielson, R. E. (1974) "The Visible Spectrum of Uranus,"
    Astrophys. Journ. (Letters), Vol. 192, L107 - L110.

Danielson, R. E. (1975) "The Structure of the Atmosphere of
    Uranus," in "The Atmosphere of Uranus," Ed. D. M. Hunten,
    NASA Ames Workshop Proceedings.

Danielson, R. E., Cochran, W. D., Wannier, P. G., and Light,
    E. S. (1977) "A Saturation Model of the Atmosphere of Uranus"
    Icarus, Vol. 31, 97 - 109.

Dunham, D. W. (1971) "The Motions of the Satellites of Uranus,"
    Ph. D. Thesis, Yale University.

Dunham, E., Elliot, J. L., and Gierasch, P. J. (1980) "The Upper
    Atmosphere of Uranus: Mean Temperature and Temperature
    Variations," Astrophys. Journ., Vol. 235, 274 - 284.

Fink, U., Benner, D. C., and Dick, K. A. (1977) "Band Model
    Analysis of Laboratory Methane Absorption Spectra from
    4500 to 10,500Å," J. Quant. Spectrosc. Radiat. Transfer.,
    Vol. 18, 447 - 457.

Fink, U., and Larson, H. P. (1979) "The Infrared Spectra of
    Uranus, Neptune and Titan from 0.8 to 2.5 Microns,"
    Astrophys. Journ., Vol. 233, 1021 - 1040.

Gillett, F. C., and Rieke, G. H. (1977) "5 - 20 Micron Observation
    of Uranus and Neptune," Astrophys. Journ. (Letters), Vol. 218
    L141 - L144.

Giver, L. P. (1978) "Intensity Measurements of the CH_4 Bands in
    the Region 4350Å to 10,600Å," J. Quant. Spectrosc. Radiat.
    Transfer, Vol. 19, 311 - 322.

Gulkis, S. (1975) "Microwave Radiometry and Implications," in
    "The Atmosphere of Uranus," Ed. D. M. Hunten, NASA Ames
    Workshop Proceedings.

Gulkis, S., Janssen, M. A., and Olsen, E. T. (1978) "Evidence
    for the Depletion of Ammonia in the Uranus Atmosphere,"
    Icarus, Vol. 34, 10 - 19

Hubbard, W. B., and MacFarlane, J. J. (1980) "Structure and
    Evolution of Uranus and Neptune," J. Geophys. Res., Vol. 85,

Hubbard, W. B., and MacFarlane, J.J. (1981) "Theoretical Predictions of Deuterium Abundances in the Jovian Planets," preprint.

Hunten, D. M. (1981) "Deep Atmospheric Temperatures for Uranus and Neptune," Private communication, unpublished.

Joyce, R. R., Pilcher, C. B., Cruikshank, D. P., and Morrison, D. (1977) "Evidence for Weather on Neptune I," Astrophys. Journ., Vol. 214, 657 - 662.

Klein, M. J., and Turanago, J. A. (1978) "Evidence of an Increase in the Microwave Brightness Temperature of Uranus," Astrophys. Journ. (Letters), Vol. 224, L31 - L34.

Klepczynski, W. J., Seidelmann, P. K., and Duncombe, R. L. (1970) "The Masses of Saturn and Uranus," Astron. Journ., Vol. 75, 739 - 742.

Lutz, B. L., Owen, Tobias, and Cess, R. D. (1976) "Laboratory Band Strengths of Methane and their Application to the Atmospheres of Jupiter, Saturn, Uranus, Neptune, and Titon," Astrophys. Journ., Vol. 203, 541 - 551.

Macy, W. W., and Sinton, W. M. (1977) "Detection of Methane and Ethane Emission on Neptune but not on Uranus," Astrophys. Journ. (Letters), Vol. 218, L79 - L81.

Macy, W. W. (1979) "On the Clouds of Uranus," Icarus, Vol. 40, 213 - 222.

Macy, W. W., Sinton, W. M., and Beichman, C. A. (1980) "Five-Micrometer Measurements of Uranus and Neptune," Icarus, Vol. 42, 68 - 70.

McElroy, M. B. (1973) "The Ionosphere of the Major Planets," Space Science Rev., Vol. 14, 460 - 473.

Michalsky, J. J., and Stokes, R. A. (1977) "Wholedisk Polarization Measurements of Uranus at Visible Wavelengths," Astrophys. Journ. (Letters), Vol. 213, L135 - L137.

Morrison, D., and Cruikshank, D. P. (1973) "Temperature of Uranus and Neptune at 2.4 Microns," Astrophys. Journ., Vol. 179, 329 - 331.

Pilcher, C. B., Morgan, J. S., Macy, W. W., and Kunkle, T. D.
    (1979) "Methane Band Limb-Brightening on Uranus,"
    Icarus, Vol. 39, 54 - 64.

Price, M. J., and Franz, O. G. (1979) "Uranus: Disk Structure
    within the 7300Å Methane Band," Icarus, Vol. 39, 459 - 472.

Prinn, R. G., and Lewis, J. S. (1973) "Uranus Atmosphere: Struc-
    ture and Composition," Astrophys. Journ., Vol. 179, 333 -
    342.

Savage, B. D., Cochran, W. D., and Wesselius, P. R. (1980)
    "Ultraviolet Albedos of Uranus and Neptune," Astrophys.
    Journ., Vol. 237, 627 - 632.

Silvaggio, P. M. (1977) "Experimental Determination of Molecular
    Absorption Coefficients for Methane and Ammonia at Low
    Temperatures and Model Atmospheres for the Major Planets,"
    Ph. D. Thesis, Cornell University.

Sinton, W. M. (1972) "Limb and Polar Brightening of Uranus at
    8870Å," Astrophys, Journ. (Letters), Vol. 176, L131 - L133.

Smith, W. H., Macy, W. W., and Pilcher, C. B. (1980) "Measure-
    ments of the H_2 4 - 0 Quadrupole Bands of Uranus and Neptune
    Icarus, Vol. 43, 153 - 160.

Stier, M. T., Traub, W. A., Fozio, G. G., Wright, E. L., and Low,
    F. J. (1978) "For Infrared Observations of Uranus, Neptune,
    and Ceres," Astrophys. Journ., Vol. 226, 347 - 349.

Trafton, L. (1967) "Model Atmospheres of the Major Planets,"
    Astrophys. Journ., Vol. 147, 765 - 781.

Trafton, L. (1972) "On the Methane Opacity for Uranus and Neptune
    Astrophys. Journ. (Letters), Vol. 172, L117 - L120.

Trafton, L. (1976) "The Aerosol Distribution in Uranus' Atmosphe:
    Interpretation of the Hydrogen Spectrum," Astrophys, Journ.
    Vol. 207, 1007 - 1024.

Trafton, L. and Ramsay, D. A. (1980) "The D/H Ratio in the Atmos-
    phere of Uranus: Detection of the $R_5(1)$ Line of HD," Icaru:
    Vol. 41, 423 - 429.

Trauger, J. T., Roesler, F. L., and Mickelson, M. (1977) "The D/H Ratio on Jupiter, Saturn, and Uranus Based on New HD and H_2 Data," Bull. Amer. Astron. Soc., Vol. 9, 516.

Wallace, L. (1975) "On the Thermal Structure of Uranus," Icarus, Vol. 25, 538 - 544.

Wallace, L., (1980) "The Structure of the Uranus Atmosphere," Icarus, Vol. 43, 231 - 259.

'ounkin, R. L. (1970) "Spectrophotometry of the Moon, Mars and Uranus," Dissertation, University of California, Los Angeles.

IMAGING OF URANUS AND NEPTUNE

Bradford A. Smith

and

Harold J. Reitsema

Department of Planetary Sciences

and

Lunar and Planetary Laboratory

University of Arizona

Tucson, Arizona, U.S.A.

ATMOSPHERES

Observations of the minute disks of Uranus and Neptune from the surface of the earth present a major challenge to any observatory site and require the most advanced techniques in optical imaging instrumentation. Less than 4 arcsec across, the disk of Uranus would fit within the Great Red Spot of Jupiter in the focal plane of a terrestrial telescope; the smaller disk of Neptune, less than 2.5 arcsec in diameter, is scarcely more than half again that of Ganymede, the largest of the Galilean satellites. At present, both planets are situated at far southerly declinations, making them even more difficult objects for Northern Hemisphere observatories.

Both Uranus and Neptune exhibit faint bluish-green disks, a consequence both of their great distance from the sun and of strong absorptions of the longer spectral wavelengths by methane in the atmosphere above their cloud tops. Visual observers (see, for example, Alexander, 1965) have occasionally depicted bands and other low-contrast markings on Uranus and Neptune, but these are certainly suspect. During the mid-1960's, C.W. Tombaugh and one of us (B.A.S.) made photographic composites of several dozen selected images of Uranus in both blue and red light; we were unable to see anything more than a featureless, limb-darkened disk. In 1970 the balloon-borne telescope, Stratoscope II, acquired several images of Uranus in visual light with a resolution of 0.2 arcsec (Danielson el al., 1972); although some investigators suggested evidence for vague, dusky markings, the cases presented were anything but convincing. Several attempts have been made recently to employ speckle image interferometry as a means for investigating the tiny disks of these distant planets. Although the prospects appear to be promising, no useful results have been obtained to date (R. Goody, 1981).

We conclude that the disks of Uranus and Neptune are probably featureless in visible light, i.e., they do not show the same type of banded, atmospheric structure which is so familiar a characteristic of the images of Jupiter and Saturn. This banded pattern, a result of strong zonal (east-west) currents in the atmospheres of these rapidly rotating planets, is made visible by the convective overturning of their atmospheres and the corresponding production of multi-hued condensibles at atmospheric levels which are observable in the visual region of the spectrum. The atmospheric circulation of Uranus and Neptune might well resemble that of Jupiter and Saturn, but the condensibles are either more homogeneous in their optical properties or, more likely, form at levels which are too deep to be seen in visible light.

During the mid-1970's, prototype versions of a new type of optical detector became available to the scientific community. This solid-state array detector is known as a charge-coupled device or CCD, and among its many superior characteristics are high quantum efficiency in the near infrared, coupled with very low readout noise. In April 1976 a 400x400-element CCD was placed at the Cassegrainian focus of the Catalina Station 154-cm reflector (University of Arizona Observatories) in a series of tests of its applicability for astronomical research (Smith, 1977). Included in a list of selected objects was the planet Uranus. Among the optical filters employed was one with a narrow transmission band centered on the 8900A absorption band of methane. The absorption within this band by methane in the Uranus atmosphere is so great that the average albedo over the disk of the planet is less than 0.02. The images, however, showed a non-uniform disk with pronounced limb brightening, which tended to be stronger toward the

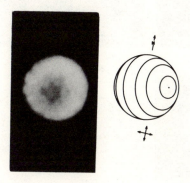

Fig. 1. CCD image of Uranus in the
8900A methane band taken in 1976.

north polar regions (Fig. 1). Although this early CCD had a
relatively high level of read-out noise, the non-uniformity of
brightness over the 3.9 arcsec disk is readily apparent. The limb
brightening must be due to an optically thin layer of particulates
at such great height in the Uranus atmosphere that back-scattering
occurs before individual photons can be absorbed by atmospheric
methane. Because the background of the methane-absorbing
atmosphere of Uranus is so dark, hazes of extremely low optical
thickness can be detected. Candidate materials for the high,
optically thin haze include methane ice crystals, photochemically
produced organic particulates and dust of exogenic origin. The
relative infrequency with which superior image quality occurs at
even a good site has thus far prevented us from obtaining the time
coverage necessary to look for rotational effects. However, the
possibility remains for an observer with dedicated instrumentation
and ample observing time at a good site.

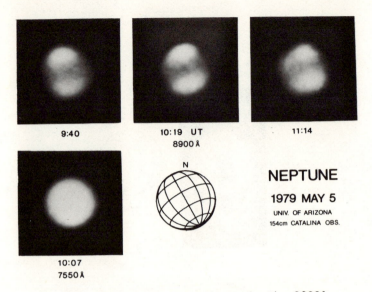

Fig. 2. CCD images of Neptune in the 8900A
methane band and continuum (7550A).

In May 1979, with a better CCD and better image quality at
the same telescope, we observed Neptune through a set of filters
centered on various methane absorption bands from 7260A to 1 micron
(Smith el al., 1979). Selected images can be seen in Fig 2. As
with Uranus, a non-uniform disk is apparent, but the appearance of
Neptune is quite different from that of its inner neighbor. The
southern hemisphere shows a general brightening; the northern
mid-latitudes, however, contain a non-axisymmetric bright feature
which is observed to move eastward with the planet's direct
rotation. The shape of the 2.5 arcsec disk has been distorted by
contrast enhancement in the computer processing of the images. One
year later, these features appeared to be less conspicuous.
Whether this difference is a result of real changes in the planet's
atmosphere or merely that of looking at a different longitude is
not known.

We are continuing to work with the methane-band images of
Uranus and Neptune. Image restoration techniques are being applied
in an effort to remove the smearing effects of the terrestrial
atmosphere and the instrumentally induced point-spread function.

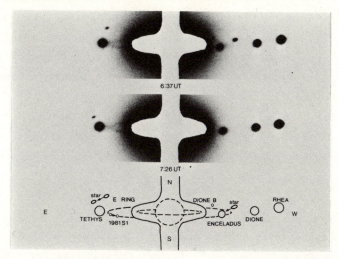

Fig. 3. Photographs of Saturn's E ring and satellites taken with the coronagraph on 1 April 1981.

Success with these image processing techniques could lead to better limb-brightening profiles, determination of the areal distribution of hazes and a rotation period for Neptune.

SATELLITES AND RINGS

The remoteness of Uranus and Neptune adds to the difficulty of observing objects very close to them, such as rings and satellites with small orbital radii. Scattered light from the relatively bright planetary disks spills over into the region of interest, i.e. the nearby space within a few planetary radii of their respective centers. However, scattered light caused by instrumental diffraction <u>can</u> be removed by a device called a focal-plane coronagraph – a modification of a technique developed by Lyot (1939). To illustrate the capability of this device and to emphasize that, even in this era of exploration by planetary spacecraft, there are still important observations to be done from the ground, Fig 3 shows recent observations of the E ring and a new satellite of Saturn made with our focal-plane coronagraph and a more classical detector - the photographic emulsion (Larson et al., 1981). The focal-plane coronagraph thus has a demonstrated capability for overcoming the long-standing problem of trying to study faint objects which are located close to bright ones (Larson and Reitsema, 1979).

The focal-plane coronagraph, when used together with a CCD, can be applied to the search for close-in satellites of Uranus and Neptune and to make optical observations of the known rings of Uranus.

OBSERVATIONS FROM SPACE

As presently scheduled, Space Telescope will be placed in orbit in January 1985 and will shortly thereafter begin imaging observations of Uranus and Neptune. The maximum resolution of the Wide Field/Planetary Camera, employing multiple imaging and image-restoration techniques, is about 40 km/au. For mean-opposition distances of Uranus and Neptune, this becomes approximately 700 km and 1200 km respectively. The respective number of resolution elements across the disks would be 70 and 40. Thus, Space Telescope will enter a realm for which no previous experience exists.

Voyager 2 is now (September 1981) on its way to Uranus and will arrive at this remote planet on 24 January 1986. Imaging resolution of atmospheric features may be as high as 20 km, while for the satellites and rings resolution will be better than 5 km and, for a few satellites, as low as 1-2 km. If Voyager 2 manages to survive an additional 3.5 years in space, it will arrive at Neptune on 24 August 1989, twelve years to the day after its launch, obtaining images of that planet, its satellites (and rings?) with resolutions comparable to those at Uranus.

REFERENCES

Alexander, A.F. O'D. 1965. The Planet Uranus (London: Faber &
    Faber).
Danielson, R.E., Tomasko, M.G. and Savage, B.D. 1972. High
    Resolution Imagery of Uranus Obtained by Stratoscope II.
    Astrophys J. 178, 887.
Goody, R.M. 1981. Private communication.
Larson, S.M. and Reitsema, H.J. 1979. A Planetary Coronagraph
    Bull. Am. Astron. Soc. 11, 558.
Larson, S.M., Fountain, J.W., Smith, B.A. and Reitsema, H.J. 1981.
    Observations of the Saturn E Ring and a New Satellite.
    Icarus, in press.
Lyot, B. 1939. A Study of the Solar Corona and Prominences Without
    Eclipses. M.N. 99, 580.
Smith, B.A. 1977. Uranus Photography in the 890-nm Absorption Band
    of Methane. Bull. Am. Astron. Soc. 9, 473.
Smith, B.A., Reitsema, H.J. and Larson, S.M. 1979. Discrete Cloud
    Features on Neptune. Bull. Am. Astron. Soc. 11, 570.

METEOROLOGY OF THE OUTER PLANETS

Garry E. Hunt
Laboratory for Planetary Atmospheres
Department of Physics and Astronomy
University College London
London WC1E 6BT

INTRODUCTION

One of the fundamental problems in atmospheric physics is concerned with providing improved weather forecasts for the Earth and predicting the future climate of our planet. However, this is a particularly difficult task since the meteorology of the Earth is affected by a wide range of physical factors – clouds, oceans, continents, polar caps, deserts and forests – all of which respond differently to the solar energy which drives the weather systems. In addition, external factors, such as the changing energy from the sun, are thought to strongly influence the Earth's meteorology. To provide a broader understanding of atmospheric phenomena, detailed studies of planetary atmospheres have been conducted, since they provide a unique opportunity to investigate these basic problems in fluid dynamics under boundary conditions which are quite different from those found on the Earth.

The giant outer planets provide, at first sight, a more varied set of objects, since they are huge, rapidly rotating, low density planets with optically reducing atmospheres (Belton (1981), Goody (1981), Kondratyev and Hunt (1981)). The giant planets Jupiter, Saturn and Neptune possess internal heat sources (MacFarlane and Hubbard (1981)) while the present observations of Uranus suggest that it may be in radiative equilibrium. The effects of internal heating as an additional driving mechanism to differential solar heating for a planetary atmosphere are not fully understood.

In order to provide insight into this fundamental problem, we discuss in this paper the meteorologies of the atmospheres of the major planets. The detailed observations of Jupiter and Saturn by the Pioneer 11 and Voyager spacecraft now make it possible to provide quantitative discussions of the motions of these planetary

atmospheres (see, for example, Hunt (1981)). We therefore first
review our basic knowledge of Jupiter and Saturn and then discuss
our current understanding of Uranus and Neptune.

THE METEOROLOGIES OF JUPITER AND SATURN

For more than 300 years, observations of large-scale cloud
features have provided the basic information on the gross
characteristics of the atmospheres of Jupiter and Saturn (see, for
example, Peek (1958), Alexander (1962) and Smith and Hunt (1976)).
The visible appearance of Jupiter is one of alternating cloud bands
of differing colours separated by jet streams. Superimposed upon
these cloud systems are large-scale features, such as the Great Red
Spot and the three white ovals, which appear to have lifetimes
varying from decades to centuries (Smith et al(1979a,b)). Saturn
is in some ways similar to Jupiter. Although the banded structure
is clearly seen, there is a marked hemispherical asymmetry in the
clarity of the visible features due to possible seasonal changes in
the upper atmospheric haze layers. Furthermore, there are
considerably less large-scale cloud features to be seen in the
Saturnian atmosphere (Smith et al (1981)).

Unlike the meteorological systems of the terrestrial planets,
the weather systems of these planets are not solely driven by
differential solar heating, since they both possess strong internal
heat sources. Jupiter emits 1.67 ± 0.13 times the energy it
receives from the sun, while the equivalent factor for Saturn is
$\sim 2.8 \pm 0.9$. Consequently, the meteorologies of these planets are
influenced by the two energy sources and by the strong rotation.
All the cloud velocities on Jupiter are referenced to the System III
period of $9^h 55^m 29.711^s$ and on Saturn to the System III period of
$10^h 39.9 \pm 0.5^m$.

On a large scale, there is little, if any, pole to equator
energy transfer at the level of the visible clouds, which is a
major difference between the Earth and the giant planets. The
Pioneer 11 measurements of Ingersoll et al (1976) have shown that
on Jupiter the difference between the equator and pole temperatures
is no more than 3K. The belt/zone structure terminates at $\pm 45^o$.

Although Saturn possesses similarly small temperature gradients, at
the time of the Voyager encounter the temperatures at the 209mb level
were noticeably warmer in the southern, compared with the northern,
hemisphere of the planet (Hanel et al (1981)). This is a seasonal
effect caused by Saturn's 27° obliquity. The radiative time constant
at this atmospheric level is approximately equal to the Saturnian
year (Cess and Caldwell (1979)), so that the thermal response is
expected to lag behind the seasonal forcing by about a quarter of
the seasonal cycle, resulting in the observed cold northern
hemisphere. On Saturn, the light and dark cloud bands also appear
to extend to latitudes of about 70° in each hemisphere.

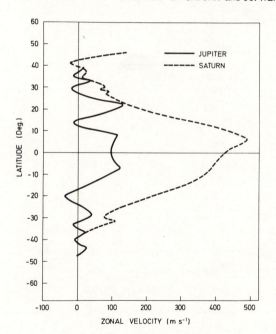

ZONAL VELOCITY vs LATITUDE on SATURN and JUPITER

In figure 1, we
have compared the
zonal velocities
of Jupiter and
Saturn, obtained
by tracking cloud
elements between
specific frames
of Voyager images.
The most notice-
able differences
between these
profiles are the
presence of the
broad, strong
equatorial jets
in the atmosphere
of Saturn, and
the virtual
absence of
easterly jets.
Retrograde jets
have been found

Fig 1 - Comparison of the zonal velocity of
        Jupiter (Voyager 1; February 1979)
with that of Saturn (Voyager 1, November 1980).

in the Saturn atmosphere at planetocentric latitudes of $\sim$41, 58,78°N

and 41 and 58°S. These major differences between the zonal wind
profiles must reflect differences in the internal structure and
driving mechanisms of these atmospheres.

Studies of the eddy momentum flux by Beebe et al (1980),
Ingersoll et al (1980) and Hunt (1981) have provided some insight
into possible driving mechanisms which are summarised in figure 2.
Available potential energy for the atmosphere is generated by the
buoyancy of the cloud systems, which is transferred into kinetic
energy by the planetary rotations. However, we find for Jupiter
that the main motions are being driven by the conversion of eddy

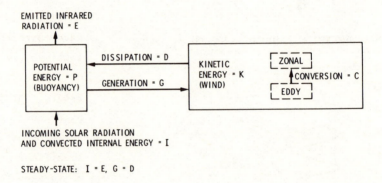

Fig 2 - A schematic representation of the energy balance
        mechanisms for the generation of the motions in
planetary atmospheres

kinetic energy into zonal mean kinetic energy as in the Earth's
atmosphere (Holton (1972)). The rate of conversion of eddy kinetic
energy into mean kinetic energy is in the range 1.5 to 3 $W.m^2$ for a
layer 2.5 bars deep. The time constant for resupply of zonal mean
kinetic energy by the eddies is in the range of 2-4 months, which is
less than the interval between the Voyager encounters. Clearly,
this situation cannot proceed unbounded, so that some larger
atmospheric system must therefore owe its existence to the extraction
of eddy momentum from the flow in which it is embedded. It is
possible that the Great Red Spot and the white ovals behave in this
way. The zonal wind profiles shown in figure 1 illustrate

essentially no correlation with the banded structure of either
Jupiter or Saturn (Ingersoll et al (1980), Hunt (1981), Smith et al
(1981)). This suggests that the visible clouds are controlled by the
radiative/dynamical properties within the upper troposphere and
clouds rather than in the deep atmosphere associated with the jets.

The rate of conversion is more than 10% of the total infrared
heat flux for Jupiter in contrast to 0.1% for the Earth. This 100
fold difference suggests that the thermomechanical energy cycles are
very different on the two planets. It is certainly possible that
the zonal flux extends much deeper than the eddies and therefore is
affected on a timescale much longer than four months. The stability
of the Jovian (and Saturnian) jets over time periods of centuries,
which is in complete contrast to the rapidly varying visible
appearance of the planet, is consistent with the assumption that the
jets are associated with the deeper layers of the atmosphere.

To explain the differences in the zonal profiles of the
planets shown in figure 1, we note that for Jupiter internal
heating supplies less energy to the atmosphere than solar heating.
For Saturn, however, internal heating supplies roughly twice the
energy than is received from solar heating. This internal heating
would tend to increase the relative importance of small scale
convection on Saturn. In addition, Saturn has a lower value for
gravity, which leads to a larger atmospheric scale height, and a
corresponding larger scale of convective motions. This would then
tend to increase the amount of available potential energy of the
system, which, on a rapidly rotating planet, would produce the
observed stronger motions (figures 1 and 2). It is apparent therefore that the energy cycle for these planets provides the basic
information to explain the observed flows, which should be taken
into account when we now address Uranus and Neptune, where
considerably less observational information is available.

THE METEOROLOGIES OF URANUS AND NEPTUNE

Uranus is the first planet so far from Earth that dynamical
activity, manifested by clouds, cannot be easily observed by ground

based observations. Consequently, we have no direct knowledge about
atmospheric motions. Our knowledge about the structure of the cloud
layers, which would act as a tracer of any motions, is itself very
limited. While methane cloud layers could form in the atmosphere
(Belton (1981)), it is also possible that any structure within this
cloud would be obscured from view by extensive aerosol layers. At
deeper levels, layers of NH_3 cloud and NH_4SH clouds are expected.
The observed changes in the microwave spectrum of Uranus by Klein and
Turegano (1978) are possibly due to variations in the structure of
these deep clouds.

Both Uranus and Neptune show long term brightness changes in
the visible portion of the spectrum. Lockwood and Thompson (1978)
have found a 1% increase in the absorption in the 6190 and 7261Å
CH_4 bands over a two month interval. Observations from 1953 to 1965
show that the B magnitude of Neptune increased from 0.02 mag., while
that for Uranus decreased from 0.04 mag. Then from 1972 to 1976,
Neptune brightened by 0.007 mag/year, while the brightness of Uranus
changed by 0.012 mag/year. These changes are certainly the result
of variations in the solar UV flux, which then alters the albedo of
the planet through the corresponding photochemical and constituent
changes to the upper atmosphere and clouds. This suggests that the
radiative budget of these planets, which must be important in the
general circulation of these planetary atmospheres, will possess
some significant variations, and thus influence the motions.

While the images of Jupiter and Saturn show definite detail
of the light/dark cloud structures, there is little to suggest that
Uranus and Neptune have a banded appearance. Alexander (1965)
suggested that when the equatorial regions of the planet are visible
and the viewing excellent two faint belts on either side of a
bright zone are sometimes seen on Uranus, using larger telescopes.
Antoniadi thought that when he observed the planet, the belts
appeared exactly parallel to the general plane of the satellite
orbits and not tilted at about 20° as other observers had reported.
Dollfus (1970) remarked that the features he saw did not seem to be
belts. The best image currently available, taken at 890nm, shows

evidence of the bright limb of Uranus, which is suggestive of a haze
layer high in the atmosphere (see Smith (1981)). Neptune has been
more difficult to observe, and the observations by Smith (1981) show
possible evidence of cloud structure with a dark equatorial belt
separating brighter regions in each hemisphere.

It is necessary to examine the basic atmospheric properties of
the planets, which are summarised in Table 1.

TABLE 1: Basic Properties of Planetary Atmospheres

	Albedo	Effective Temp(^{O}K)	Measured Temp(^{O}K)	Adiabatic Lapse Rate (^{O}K/km)	Radiative Time Constant
Earth	0.30	256	256	~9.8	~60 days
Jupiter	0.33±0.02	106	124.9±0.3	1.9	~6 yrs
Saturn	~0.36	76±4	96.5±2.5	0.9	~6 yrs
Uranus	0.34–0.5	55–58	58±2	~1.0	~600 yrs
Neptune	0.34–0.5	43–46	~55	~1.4	~2200yrs

These tabulations indicate that, unlike the other major planets,
Uranus does not appear to possess an internal heat source to provide
an additional driving mechanism for the motions. The long radiative
time constant compared with the length of the Uranian day (Houghton,
1977; Goody, 1981) would suggest a relatively more stable atmosphere
than the other planetary atmospheres. Stone (1973, 1975) suggested
that the Richardson number may be as large as 2800, confirming this
interpretation.

Uranus does, however, possess some important differences from
the other planets. The large inclination of the axis of rotation
will mean that each hemisphere will spend a substantial period
without sunlight once each Uranian year. At the present time, the
north pole of the planet is just turning towards the Earth after
having turned away from the sun for more than 40 years. The
atmosphere will then cool substantially in the absence of meridional
heat transport.

Although Uranus has a rapid rotation period of 16.31±0.27 hrs

(Goody (1981)), the large inclination of the equator means that
during the course of one orbit all equatorial regions receive less
heat from the sun than the polar latitudes. Consequently, the poles
will be hotter than the equator, so that from the thermal wind
equation the mean zonal winds will be easterly, in contrast to the
mean westerly winds of Jupiter and Saturn.

Even in the absence of any major internal heating, it is
natural to expect that on a rapidly rotating planet certain
natural horizontal scales of motion will be generated. We may
anticipate barotropic instabilities of $\sim$1000km and baroclinic
instabilities of $\sim$600km for a mean velocity of $\sim$2ms^{-1}. The
corresponding time scales of these features would be about 40 days
and 8 days respectively. We would not necessarily expect to find
large scale convective motions in the atmosphere of Uranus. As a
consequence, we may expect the motions to be far less turbulent
than on Jupiter and Saturn, and possibly even the Earth.

Neptune may, however, have large scale motions. Joyce et al
(1977) have observed a factor of four change in the reflectance
of the planet between 1 and 4μm during the period 1975-76, while the
simultaneous measurements of Uranus showed a negligible change.
More recently, Cruikshank (private communication (1981)) has
reported similar rapid changes in the structure of Neptune. Pilcher
(1977) suggested that these changes could be accounted for by large
scale variations in the cloud structure. Since the observations
refer to the whole planetary disk, we may anticipate considerable
structure in the cloud layers when higher spatial resolution
observations are made later this decade. Certainly, it is possible
that Neptune could have strong motions. Belton et al (1981) believe
that the simultaneous presence of three periodicities of 17.73,
18.56 and 19.29 hours for the rotational period of the planet,
together with related harmonics in the J-K colour, is consistent
with zonal winds of up to 109ms^{-1}. Furthermore, the internal
heating would be dominant over the available absorbed solar energy
and therefore by the energy budgets of figure 2, leading to strong
motions.

CONCLUSIONS

The meteorology of Uranus may be quite different from the motions of the other large planets. The apparent absence of internal heating and the large obliquity of the axis of rotation on this rapidly rotating planet make Uranus quite different from its planetary neighbours. We anticipate only weak motions, generated primarily by the differential solar heating, which, at the time of the Voyager encounter in 1986, will be a maximum in the polar region. The meteorology of Uranus provides a major test of our understanding of planetary weather systems, and must be monitored by Space Telescope during the next decade to extend the high spatial Voyager observations.

ACKNOWLEDGEMENTS

This work is supported by the Science and Engineering Research Council. Contribution 97 of the Laboratory for Planetary Atmospheres.

REFERENCES

Alexander, A.F.O'D (1962) The Planet Saturn (Faber and Faber)

Alexander, A.F.O'D (1965) The Planet Uranus (Faber and Faber)

Beebe, R.F., Ingersoll, A.P., Hunt, G.E., Mitchell, J.L. and Müller, J-P (1980) Measurements of Wind Vectors, Eddy Momentum Transports and Energy Conversions in Jupiter's Atmosphere from Voyager 1 Images. Geophys. Res. Letts. 7, 1-4

Belton, M.J.S. (1981) An Introductory Review of our Present Under-standing of the Structure and Composition of Uranus' Atmos-phere (this volume)

Belton, M.J.S., Wallace, L. and Howard, S. (1981) The Periods of Neptune; Evidence for Atmospheric Motions. Icarus (in press)

Cess, R.D. and Caldwell, J. (1979) A Saturnian Stratospheric Seasonal Climate Model. Icarus 38, 349-357

Dollfus, A. (1970) New Optical Measurements of the Diameters of Jupiter, Saturn, Uranus and Neptune. Icarus 12, 101-117

Goody, R. (1981) The Rotation of Uranus (this volume)

Hanel, R. et al (1981) Infrared Observations of the Saturnian System from Voyager 1. Science 212, 192

Holton, J. (1973) An Introduction to Dynamical Meteorology
    (Academic Press)

Houghton, J. (1977) The Physics of Atmospheres (CUP)

Hunt, G.E. (1981) The Atmospheres of Jupiter and Saturn. Phil.
    Trans. Roy. Soc. (in press)

Ingersoll, A.P., Munch, G., Neugebauer,G., Orton, G.S. (1976)
    Results of the Infrared Radiometer Experiment on Pioneers 10
    and 11 in Jupiter (ed.by T. Gehrels) p.197-205. University
    of Arizona Press

Ingersoll, A.P., Beebe, R.F., Mitchell, J.L., Garneau, G., Yagi, G.,
    Müller, J-P and Hunt, G.E. (1980) Interaction of Eddies and
    Mean Zonal Flow on Jupiter as Inferred from Voyager 1 and 2
    Images. J. Geophys. Res. (in press)

Joyce, R.R., Pilcher, C.B., Cruikshank, D.P, and Morrison, D. (1977)
    Evidence for Weather on Neptune I. Astrophys. J. 214, 657-
    662

Klein, M.J. and Turegano, J.A. (1978) Evidence of an Increase in
    the Microwave Brightness Temperatures of Uranus. Astrophys.
    J. 224, L31-34

Kondratyev, K.Ya and Hunt, G.E. (1981) Weather and Climate on
    Planets (Pergamon Press)

Lockwood, G.W. and Thompson, D.T. (1978) A Photometric Test of
    Rotational Periods for Uranus and Time Variations of Methane
    Band Strengths. Astrophys. J. 221, 689-693

MacFarlane, J.J. and Hubbard, W. (1981) Internal Structure of
    Uranus (this volume)

Peek, B. (1958) The Planet Jupiter (Faber and Faber)

Pilcher, C.B. (1977) Evidence for Weather on Jupiter II.
    Astrophys. J. 214, 663-666

Smith, B.A. et al (1979a) The Jupiter System through the Eyes of
    Voyager 1. Science 204, 951-972

Smith, B.A. et al (1979b) The Galilean Satellites and Jupiter :
    Voyager 2 Imaging Science Results. Science 206, 927-951

Smith, B.A. et al (1981) Encounter with Saturn:Voyager 1 Imaging
    Science Results. Science 212, 163-191

Smith, B.A. (1981) Imaging Studies of the Outer Planets (this
    volume)

Smith, B.A. and Hunt, G.E. (1976) Motions and Morphology of Clouds
    in the Atmosphere of Jupiter   in Jupiter (ed.by T. Gehrels)
    (University of Arizona Press)

Stone, P. (1973) The Dynamics of the Atmospheres of the Major
    Planets. Space Sci. Rev. 14, 444-459

Stone, P. (1975) The Atmosphere of Uranus. Icarus 24, 292-298

THE SATELLITES OF URANUS

by

Dale P. Cruikshank

Institute for Astronomy, University of Hawaii

2680 Woodlawn Drive, Honolulu, Hawaii 96822

ABSTRACT

The Uranian satellite system contains five known members, all of which are difficult to study owing to their faintness and proximity to Uranus. The photometry of these objects is not in a satisfactory state, nor is the photovisual spectrophotometry. Infrared observations are in some sense easier and more precise because Uranus itself is faint in the near infrared and interferes but little with the satellite studies. The near infrared work reveals water ice or frost on the satellite surfaces, perhaps in a very pure state. The satellites are most similar to Ganymede in terms of the strength of the ice bands, but subtle differences, now under study, may be present. The diameters and masses of the satellites, while not readily measurable directly, can be estimated from generalizations about the surface geometric albedos, assumed mean densities, and certain dynamical arguments. Most of the data and theories are consistent with bodies with radii in the range 160-520 km (similar to the larger asteroids) and albedos on the order of 0.5, consistent with ice and snow. The mean densities are probably similar to those of the icy Saturn satellites, about 1.3 g/cm^3. Uranus' satellites probably formed after the event that caused the planet to tilt to the presently observed obliquity.

INTRODUCTION

Following William Herschel's discovery of the planet Uranus in 1781, he continued to build larger and better telescopes, culminating in the great 40-foot telescope at Slough. With a telescope of 18.7 inches aperture he found the outer two satellites, U3 Titania, and U4 Oberon, in 1787, and then searched for additional satellites

for the next 14 years. Though he eventually announced the discovery of four additional satellites (in 1798), subsequent observations by others suggested that the positions and periods of
revolution given by Herschel were spurious. William Lassell, the
discoverer of Triton, observed the planet with telescopes of probably higher quality (better rejection of scattered light from the
planet), and reported in 1851 the discovery of U2 Umbriel and U1
Ariel. In a detailed examination of the circumstances of discovery,
D. Rawlins in a paper at this conference has given evidence that
Herschel should be given credit for the discovery of Umbriel in
1801 and Otto Struve credit for Ariel in 1847, though the essential
work of Lassell in firmly establishing the orbital periods is not
doubted. Additional historical analysis of the question of
satellite discoveries will be found in Alexander (1965). Sir John
Herschel suggested the names that are now in use for the first four
satellites (see Barton, 1946). Oberon and Titania are the king and
queen of the fairies in Shakespeare's *Midsummer Night's Dream*.
Ariel and Umbriel appear in Pope's *Rape of the Lock*, and Ariel is
"an airy, tricksy spirit" in Shakespeare's *Tempest*.

The last satellite discovery for Uranus was that of Miranda by
G. P. Kuiper in February, 1948. He found the innermost satellite
at magnitude 16.5 on a four-minute photographic exposure with the
82-inch McDonald Observatory telescope, at that time the largest
telescope with which a detailed survey of satellite systems had
been conducted. The same survey revealed N2 Nereid in 1949.
Kuiper (1949) named U5 Miranda after a cherub in the *Tempest*.

The orbital data and visual magnitudes of the five known
satellites are given in Table I. The orbital data are taken from
Morrison and Cruikshank (1974) who extracted them from Gondolatsch
(1965) and Whitaker and Greenberg (1973).

TABLE I.   BRIGHTNESS AND ORBITAL PROPERTIES OF THE URANIAN
            SATELLITES

	Visual Brightness (mean opposition)	Orbital Radius		Period (days)	Eccentricity	Inclination (degrees)
		$[10^3 \text{km}]$	(planetary radii)			
U5 Miranda	16.5	130	5.13	1.4135	0.017	3.4
U1 Ariel	14.4	192	7.54	2.520	0.0028	0
U2 Umbriel	15.3	267	10.5	4.144	0.0035	0
U3 Titania	14.0	438	17.2	8.706	0.0024	0
U4 Oberon	14.2	586	23.0	13.46	0.0007	0

All five satellites and Uranus are nicely shown in the composite photograph in Figure 1 provided by William Sinton (1972).

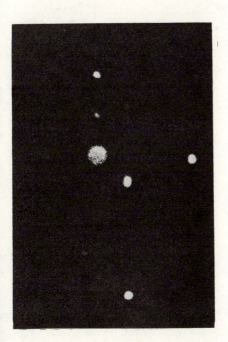
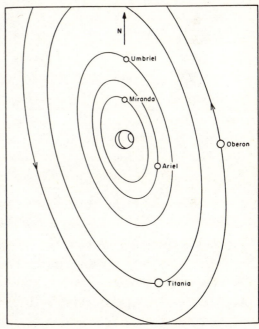

Figure 1.   The composite photograph on the left shows Uranus and the five known satellites, as identified in the guide on the right.   The Uranus image is a composite of six individual one-minute exposures on 17 March and 7 May, 1972, while the satellites were recorded on an exposure of about one hour on 17 March.   The exposures of the planet were made through a filter transmitting in the region of a strong methane band which is heavily absorbed in the Uranus spectrum.   The nonuniform density of the Uranus image is real and is related to the distribution of haze in the upper atmosphere of the planet.   From Sinton (1972).

Photometry

The earliest photometric work on the Uranian satellites was the
set of broadband measurements on Titania and Oberon by Harris
(1961) with the 82-inch McDonald reflector. The circumstances of
the observations were never published, but have to some degree been
reconstructed by Andersson (1974). New measurements of the same two
satellites by Andersson, Degewij and Zellner have just appeared
(Degewij, et al. 1980a); their values of V(1,0) are given in Table
II.

TABLE II. PHOTOMETRY OF THE URANIAN SATELLITES

Satellite	Harris (1961)	Reitsema *et al.*[c] (1978)	Degewij *et al.* (1980a,b)
U5 Miranda	3.8[a]	3.79 ± .17	—
U1 Ariel	1.7[a]	1.48 ± .15	—
U2 Umbriel	2.6[a]	2.31 ± .15	—
U3 Titania	1.3[b]	1.27 ± .14	1.27 ± .14[b]
U4 Oberon	1.5[b]	1.42 ± .15	1.52 ± .05[b]

Notes:

[a] Photographic estimate.

[b] Photoelectric measurement.

[c] Photometry from Charge-Coupled Device (CCD) images.

The magnitudes of the other three satellites in Harris' listing are
photographic estimates based on plates obtained largely by T.
Gehrels; direct photometric measurements, even with large telescopes,
are very difficult because of the proximity of Uranus. Additional
differential photometry of all five satellites was obtained by
Reitsema et al. (1978) using a charge-coupled device (CDD) yielding
images of the satellite system through four different color filters,
including the broadband V filter. The Reitsema et al. (1978) data
were referenced to the brightest object, Titania. In Table II, I
have used the Degewij et al.(1980a,b) value of V(1,0) for Titania with
its standard error of ± 0.14, and list the resulting Reitsema et
al. (1978) values from their published differential values, propa-
gating their estimated errors. Shown also are the earlier values

from Harris (1961), though no error estimates are available for
these data. It can be seen that the Degewij et al. (1980a,b) and
Reitsema et al. (1978) values for Oberon do not agree as well as
one might hope for objects as bright as these and at relatively
large distances from the planet (Titania is about 34 arcsec from
Uranus and Oberon is about 45 arcsec). They do agree within the
values of the error bars, however. The agreement with the Harris
photometry is good for Titania, Oberon, and Miranda, though Ariel
and Umbriel are not in good accord. My colleagues who are familiar
with CCD imagery tell me that photometry with these systems has not
reached its full degree of perfection, which may account for part
of the difficulty in Table II, but the differences in the Harris
values and the more recent data may in part result from the
different viewing geometry at the epochs of the two sets of observations, the more recent representing a more nearly polar view of
the satellite system. Note that the photometry summary given by
Cruikshank (1980, Table II) used values of Titania and Oberon from
Andersson's (1974) thesis. His data have been superceded by the
new analysis in Degewij et al. (1980a), though the differences in
V_o [or $V(1,0)$] are small. Table III gives a summary of the meager
information at hand.

TABLE III. PHOTOMETRIC COLORS OF THE URANIAN SATELLITES

	U–B	B–V
Harris		
Titania	0.25	0.62
Oberon	0.24	0.65
Degewij *et al.*		
Titania	0.28	0.68
Oberon	0.22	0.69
Reitsema *et al.*		

"Umbriel, Titania, and Miranda are very similar
(0.5 - 1.0 μm), while Ariel and Oberon form a
second spectral reflectance group which is
slightly less red."

Broadband photometry in the J (1.25 μm), H (1.60 μm), and K
(2.2 μm) bands can, in principle, help distinguish solar system

bodies of icy or rocky surface composition (Morrison et al. 1976;
Cruikshank 1980; Degewij et al. 1980b; Hartmann et al. 1981 and
others). The J-H and H-K colors for icy objects tend to be distinctly different from those of rocky bodies, such as asteroids and
low-albedo planetary satellites, but the colors of the Uranian
satellites fall in a portion of the J-H, H-K color diagram that is
a bit ambiguous and characterized by nearly neutral coloration.
This information, combined with the relatively neutral colors in
the photovisual spectral region, leaves a serious ambiguity in the
deduction of surface composition, though this has been clearly
resolved by spectrophotometry in the near infrared, as discussed
below. The broadband JHK photometry of the Uranian satellites has
been obtained by Cruikshank et al. (1977), Cruikshank (1980), and
Nicholson and Jones (1980). The spectrophotometry is somewhat more
interesting than the broadband photometry, and the photometry will
therefore not be discussed here in any detail.

Photovisual Spectrophotometry

In this context the term photovisual refers to data obtained
with electronic detectors in the spectral region 0.3-1.1 μm. The
first such data with significant spectral resolution obtained for
the Uranian satellites are those of Johnson et al. (1978) who used
the Oke scanner on the Hale 5-m telescope. They found that the
reflectances of Titania, Ariel, and Oberon drop significantly
shortward of 0.5 μm, with the reflectance at longer wavelengths
essentially flat for Ariel and Titania, while that of Oberon
continued to rise significantly. In contrast, the reflectance of
Umbriel was found to rise toward the violet. They also published
data for Triton which are at variance with results of similar
quality obtained by other means (Cruikshank et al 1979; Bell et al.
1979). The Johnson et al. (1978) data for Titania and Oberon show
a much steeper ultraviolet drop than is supported by the UBV
photometry given in Table III. Further, multi-filter spectrophotometry by Bell et al. (1979) of Titania and Oberon is distinctly
different from the Johnson et al. (1978) data in the sense that no

strong violet absorption is present in the former results.   It can
easily be argued that systematic errors in observation or reduction
most often result in the <u>loss</u> of light in some spectral region
rather than the addition of light, and that such errors would
produce an apparent absorption in a spectrum rather than a tendency
toward spectral neutrality.   The Bell <u>et</u> <u>al</u>. (1979) data for Titania
and Oberon, shown in Figure 2 clearly reveal a rather neutral color
throughout the spectral range covered.

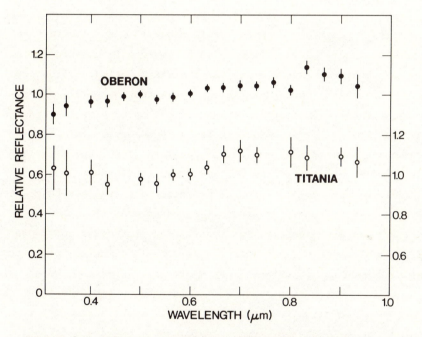

Figure 2.   Reflectance spectra of Titania and Oberon, each
normalized to unity at 0.56 µm.   The reflectance scale on the left
is that for Oberon and that on the right for Titania.   From the work
of Bell, <u>et</u> <u>al</u>. (1979), previously unpublished.

The disagreement between these two data sets, the strong coloration
in the Johnson <u>et</u> <u>al</u>. (1978) data, and the latter's disagreement
with the U–B, B–V colors, suggests that the Johnson <u>et</u> <u>al</u>. (1978)
data are flawed by systematic effects.   They were taken with a
rather small aperture of 5 arcsec at the relatively large airmass
(at least 1.36) forced by the southerly declination of Uranus.   I

suggest that differential atmospheric refraction caused the loss of
the blue and violet light in their observations, resulting in what
appears to be a strong photometric signature, and that their spectrophotometric curves are therefore spurious. For the present
discussion, the relatively neutral reflectance of Titania and
Oberon is regarded as fairly well established, while no satisfactory
data on the other satellites as yet have been obtained.

Near-Infrared Spectrophotometry

I published the first infrared spectrophotometry of two Uranian
satellites, Titania and Oberon, (Cruikshank 1980) and showed that
their reflectances between 1.40 and 2.50 μm reveal the strong absorptions (at ∿1.5, 2.0, and 2.4 μm) characteristic of <u>water frost</u> or
<u>ice</u>. In the case of both satellites, the strengths of the absorp-
tion at ∿2.0 and 2.4 μm are less than in the spectra of the rings of
Saturn and are quite comparable to those in the spectrum of Jupiter's
Ganymede, as studied in great detail by Clark (1980) and Clark and
McCord (1980a,b). Apart from the H_2O found in comets, the new
measurements represent the greatest distance from the sun that water
has been found in the solar system. Spectrophotometry, though
crude, for Ariel and Umbriel was reported by Cruikshank and Brown
(1981), in which it was shown that these satellites also have water
frost or ice on their surfaces. R. H. Brown is studying the spectra
of the Uranian satellites in detail as part of a Ph.D. thesis at the
University of Hawaii. In the course of this work, he and I have
obtained, in 1981, data of superior quality to those published in
the two papers referenced. The significant improvement in the data
has been achieved by the use of improved detectors and a much more
satisfactory circular-variable interference filter (CVF) which
provides the wavelength separation in the InSb photometric system
used. The CVF used in the early work had a bandpass of ∿1%, while
the new filter has 5% and overall higher transmission.

To first order, the new 1981 data confirm the conclusions
reached from the earlier studies, that is absorptions of H_2O ice or
frost dominate the near infrared spectrum with very strong absorptions at the appropriate wavelengths. In my 1981 paper with Mr.

Brown we suggested that the spectrum of Umbriel shows ice bands of less strength than those of the other three satellites. While some differences among the new spectra are evident, it is not apparent that the earlier conclusion about Umbriel is borne out. The new data will permit a detailed comparison with laboratory spectra of ices and mixtures of ices and mineral contaminants, much in the way that Clark (1980) and Clark and McCord (1980a,b) have done for the Galilean satellites and the rings of Saturn, and this will be substantially the direction of Mr. Brown's effort in the next year. A preliminary conclusion from the new data is that the spectra of the satellites do indeed differ from one another in subtle ways not discernable in the rougher data heretofore available, but clear trends or causes of the differences are not yet understood. The interested reader is advised to watch for Mr. Brown's future publications on this subject.

In the meantime, I present here the published data in the form of Figures 3 and 4. Figure 3 shows the spectrum of Titania (points with error bars) in two representations. At the top of the figure the solid line is the spectrum of Saturn's rings and at the bottom it is the spectrum of Ganymede, in both cases provided by Dr. R. Clark. The Titania points are properly normalized to the respective comparison spectra by averaging over several points near 2.2 µm. This figure affirms the earlier statement that the Titania profile more closely resembles that of Ganymede than the rings of Saturn in terms of the depths of the distinct ice absorptions at $\sim$2 and 2.4 µm. The new 1981 data noted above were obtained at Mauna Kea Observatory where the smaller airmass and drier sky permitted precise photometry right through the 1.9-µm telluric water band, so the new data are not interrupted mid-spectrum as are the old, and the profile of the ice band at $\sim$2 µm is well shown.

Published spectra for all four bright satellites of Uranus are given in Figure 4 with the Ganymede spectrum given for comparison in all cases. The spectra are arranged top to bottom in order of increasing apparent depth of the ice bands, though the new 1981 data may disrupt this ordering when they are fully analyzed.

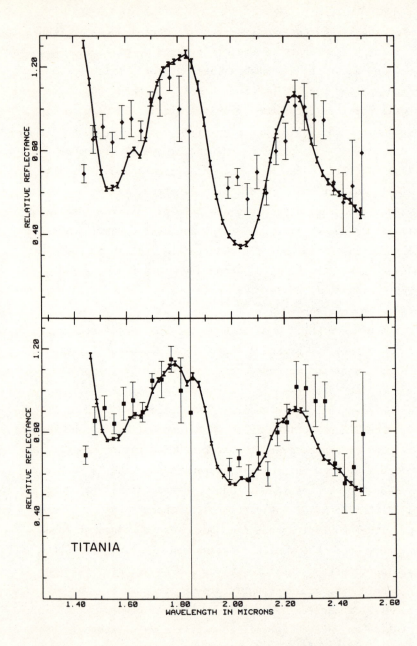

Figure 3.   Spectrum of Titania (points with one-sigma error bars) compared with the spectrum of Saturn's rings (top) and the spectrum of Ganymede (bottom).   Strong absorptions in the spectra of the rings, Ganymede, and Titania, are due to H_2O ice or frost. From Cruikshank (1980).

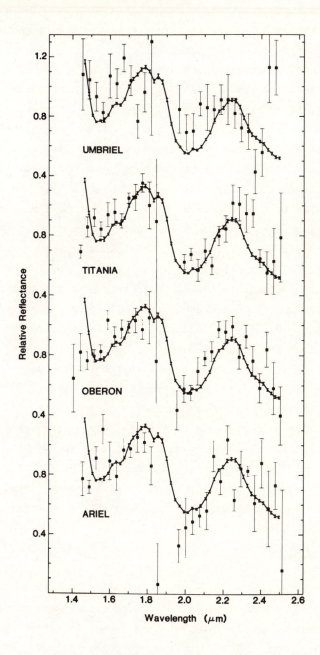

Figure 4. Spectra of the four outer Uranian satellites (points with one-sigma error bars) compared with the spectrum of Ganymede (solid lines). From Cruikshank and Brown (1981).

The appearance of water frost in a relatively pure state
(freedom from mineral contaminants) would support the apparent
neutrality of the photovisual and near infrared spectra noted
earlier from the broadband and photovisual spectrophotometry.
Still, the ice bands are less deep than on Saturn's rings (regarded
as quite pure by Clark and McCord, 1980b), and we have seen in
Voyager pictures of Ganymede that its surface is quite varied and
has evidence of geological structures suggesting materials in
addition to ice on its surface. Further, the fact that the ice
bands are weaker on Ganymede than on the rings suggest a contami-
nant filling in the bands. Clark (1980) concludes from an extensive
laboratory study of ices and their contaminants that such mineral
contaminants as occur in the frost in Saturn's rings are not on the
surfaces of the particles but a few micrometers to millimeters
under the uppermost surface where they can lower the continuum
level a small amount but not strongly affect the band depths. In
contrast, mineral grains on Ganymede occur at the surface of the
fine-grained frost layer, where they affect both the intensity of
the continuum and the strength of the absorption bands. On the
other hand, Ganymede shows a strong violet absorption in its spectrum, probably as a consequence of these mineral contaminants, and
we have argued above that the Uranian satellites do not have the
violet absorption. Clearly there is a lot here for Mr. Brown to
explain in his study of these interesting little satellites, and it
seems clear that his new data set will provide the necessary
fundamental information for the investigation.

Physical Properties

Apart from the compositional conclusions derived from the
information given above, there is no direct information on the
physical properties of the Uranian satellites. We cannot yet
measure their masses, their diameters, geometric albedos, temperatures, or figures. Some of this information can be pieced together from a string of inferences, as we shall see, but first I
will review what is known about the masses of the satellites based

on relatively recent theoretical discussions and on new observational programs in progress.

Greenberg (1975, 1976, 1979) has given the most recent thorough
summary of the situation with regard to the satellite masses and has
reviewed all the earlier literature. Greenberg analyzes the Laplace
relation among the satellites and shows in a second-order perturbation theory that the products of the masses of certain satellite
combinations can be found to establish certain constraints. The
results are, for masses denoted $m_1 \ldots m_5$ for the satellites Ariel
through Miranda,

$$m_1 \cdot m_2 \lesssim 10^{-9}$$
$$m_5 \cdot m_1 \lesssim 5 \times 10^{-12}$$
$$m_5 \cdot m_2 \lesssim 6 \times 10^{-12}$$

in units of the mass of Uranus. Dr. Rattier at this conference has
used new observations of Miranda's precession to derive

$$m_1 \cdot m_2 = (1.10 \pm .25) \times 10^{-10}$$

and finds $m_3 = 1.5 \times 10^{-4}$ from new data based on 200 plates of
Miranda. Also at this conference, Dr. Ferraz-Mello has quoted the
value $m_3 = (1.47 \pm .10) \times 10^{-4}$ from the 1981 work of Veillet. It
has been gratifying to learn at this conference that there is
renewed interest in determining the masses of the satellites from
telescopic observations over a long time base and that both the
observational and theoretical work are now going on at a rapid pace.

We can approach the problem of the masses of the satellites
from another viewpoint as we try to understand simultaneously the
radii, mean densities, and surface geometric albedos of these
bodies. The radius, r, of a Uranian satellite is related to its
geometric albedo, p_V, and observed stellar magnitude, $V(1,0)$, by

$$5 \log r = V_\odot - V(1,0) - 2.5 \log p_V$$

where $V_\odot$ is the stellar magnitude of the sun, -26.77. The mass is
$m = \frac{4}{3} \pi r^3 \bar{\rho}$ where $\bar{\rho}$ is the mean density in g/cm^3.

We have established the presence of water ice or frost on the
surface of four of the satellites and can presume that it also
exists on Miranda. We have also given evidence that the spectral

reflectances are rather neutral, both of which pieces of information
suggest a relatively high geometric albedo comparable to snow (about
0.6 or greater). The close similarity to Ganymede tends to argue
for a similar geometric albedo (0.43) to that satellite. In all,
the best estimate of p_V for the Uranian satellites is in the range
0.4-0.6. Further, studies of the icy satellites of Jupiter and
Saturn (particularly the latter) show mean densities in the range
1-2 g/cm^3. A reasonable estimate for the icy satellites of Uranus
might be 1.3 g/cm^3, the same as Saturn's Rhea.

In Table IV, I give a sample set of calculations of the radii
of the Uranian satellites for different values of the geometric
albedo. For the purposes of further discussion here we can assume
$p_V = 0.5$, giving radii in the range 160-520 km.

TABLE IV. RADII (KM) OF THE URANIAN SATELLITES

	p_V					
	0.1	0.2	0.4	0.5	0.6	0.8
U5 Miranda	360	260	180	160	150	130
U1 Ariel	1050	740	520	470	430	370
U2 Umbriel	710	500	360	320	290	250
U3 Titania	1150	810	580	520	470	410
U4 Oberon	1020	720	510	460	420	360

In the verbal presentation of this paper at Bath, I suggested that
the spectrophotometric evidence favored a slightly lower value of p_V
for Umbriel, vis. 0.3. The new observations obtained since the Bath
conference do not support that contention, and there is presently no
evidence for differences among the four satellites so far observed
in the near infrared. We thus use $p_V = 0.5$ for all of the objects.

If we then assume $\bar{\rho}$ 1.3 g/cm^3, we can calculate the masses
corresponding to the diameters derived above and examine them for
compliance with the Greenberg criteria and the new derivations
referenced above. It will be seen in Table V that the Greenberg
mass constraints can easily be met for all values of the radius
derived from $0.4 \leqslant p_V \leqslant 0.6$.

TABLE V. MASSES OF THE URANIAN SATELLITES[a]

	p_V		
	0.4	0.5	0.6
U5 Miranda	0.37	0.26	0.21
U1 Ariel	8.8	6.5	5.0
U2 Umbriel	2.9	2.1	1.5
U3 Titania	12.3	8.8	6.5
U4 Oberon	8.3	6.1	4.7

[a] Assumes $\bar{\rho}$ = 1.3 g/cm^3. All masses in units 10^{-6} mass of Uranus.

Dr. Rattier's estimate of m_3 is not within the range found from the considerations discussed, requiring a rather large mean density and lower albedo, though his value of $m_1 \cdot m_2$ can be met with only a slightly lower albedo and/or higher mean density than the preferred range noted. Clearly, additional work is needed, and the assumptions upon which this simple analysis is based need refinement, but we see that most of the constraints can be met for plausible ranges of the basic physical parameters. We also see that the basic discovery of water ice on these satellites is very important in establishing them as objects of rather high albedo, hence relatively small size. For p_V = 0.5, Table IV shows that the Uranian satellites are comparable in radius to the largest asteroids and the satellites of Saturn (except Titan).

Origin of the Uranian Satellites

The peculiar dynamical state of Uranus and its satellites has been the focus of most studies of the origin of this system to date. Uranus has an obliquity of about 98° and the satellite system is very regular (mostly circular orbits of very low inclination) and lying in the planet's equatorial plane. Any theory of the origin of the satellites must take into account the mechanism and time scale for the tilting of the planet itself, the presumption being that Uranus condensed with its spin axis nearly perpendicular to the ecliptic, followed by a tilting event. Singer

(1975) has considered several scenarios and favors the formation or acquisition of the satellites after the planet tilted. He suggests the tidal capture of a body having 5–10% of the planet's mass, after which the body broke up and some 1% of it remains in the form of the five observed (and perhaps some unobserved) satellites.

From the point of view of composition some remarks can be made. The rings of Uranus appear to be composed of dark particles (Sinton 1977; Smith 1977), forming yet another contrast with the system of Saturn's satellites and rings. The total mass of the Uranian rings is probably on the order of 10^{-12} the mass of the planet, and thus far less than that of even Miranda. The satellites appear to be composed of volatile material, water ice. This difference in the composition of ring and satellite material may support Singer's hypothesis and may represent some degree of spatial differentiation of material from a disrupted body that provided the matter from which the satellites originated. Dr. Podolak has presented other information and ideas on the origin of Uranus and the satellites at this conference.

The Future

Continued improvements in observational equipment will permit further ground-based studies of the satellites and rings of Uranus, perhaps giving a clearer understanding of the interrelation between the two systems, both dynamically and chemically. Additional satellites may be found, particularly closer to the planet and perhaps intimately associated with specific rings in the nine-component (at least) system.

The Voyager 2 flyby of the Uranus system in 1986 will afford brief and imperfect glimpses of some of the satellites, particularly Miranda and Ariel, but the approach and viewing geometry are not optimum for a satellite-centered mission as a consequence of the planet's tilt. Still, the first close look at any components of the Uranian system is likely to help us understand the aggregate, and is anticipated with the greatest enthusiasm.

ACKNOWLEDGEMENT

This research and travel to the Bath conference were supported in part by NASA Grant NGL 12001-057. I thank R. H. Brown for his contribution to this work.

REFERENCES

ALEXANDER, A. F. O'd. 1965. *The Planet Uranus* (New York: American Elsevier) 300 pp.

ANDERSSON, L. E. 1974. A photometric study of Pluto and satellites of the Jovian outer planets. Ph.D. thesis, Univ. of Indiana, Bloomington.

BARTON, S. G. 1946. The names of the satellites. *Pop. Astron.* 54, 122-131.

BELL, J. F., CLARK, R. N., McCORD, T. B., and CRUIKSHANK, D. P. 1979. Reflection spectra of Pluto and three distant satellites. *Bull. Amer. Astron. Soc.* 11, 572 (Abstr.).

CLARK, R. N. 1980. Ganymede, Europa, Callisto, and Saturn's rings: Compositional analysis from reflectance spectroscopy. *Icarus* 44, 388-409.

CLARK, R. N., and McCORD, T. B. 1980a. The Galilean satellites: New near-infrared spectral reflectance measurements (0.65 - 2.5 μm) and a 0.325 - 5 μm summary. *Icarus* 41, 323-339.

CLARK, R. N., and McCORD, T. B. 1980b. The rings of Saturn: New near-infrared reflectance measurements and a 0.326 - 4.08 μm summary. *Icarus* 43, 161-168.

CRUIKSHANK, D. P. 1980. Near-infrared studies of the satellites of Saturn and Uranus. *Icarus* 41, 246-258.

CRUIKSHANK, D. P., and BROWN, R. H. 1981. The Uranian satellites: Water ice on Ariel and Umbriel. *Icarus* 45, 607-611.

CRUIKSHANK, D. P., PILCHER, C. B., and MORRISON, D. 1977. Identification of a new class of satellites in the outer solar system. *Astrophys. J.* 40, 1006-1010.

CRUIKSHANK, D. P., STOCKTON, A., DYCK, H. M., BECKLIN, E. E., and MACY, W. Jr. 1979. The diameter and reflectance of Triton. *Icarus* 40, 104-114.

DEGEWIJ, J., ANDERSSON, L. E., and ZELLNER, B. 1980a. Photometric properties of outer planetary satellites. *Icarus* 44, 520-540.

DEGEWIJ, J., CRUIKSHANK, D. P., and HARTMANN, W. K. 1980b. Near-infrared colorimetry of J6 Himalia and S9 Phoebe: A summary of 0.3- to 2.2-μm reflectances. *Icarus* 44, 541-547.

GONDOLATSCH, F. 1965. In *Landolt-Bornstein Tables, New Series*, eds. K. H. Hellwege and H. H. Voigt, vol. 1 (Berlin: Springer-Verlag) p. 150.

GREENBERG, R. 1975. The dynamics of Uranus' satellites. *Icarus* 24. 325-332.

GREENBERG. R. 1976. The Laplace relation and the masses of Uranus' satellites. *Icarus* 29, 427-433.

GREENBERG, R. 1979. The motions of Uranus' satellites: Theory and application. In *Dynamics of the Solar System*, ed. R. L. Duncombe (Dordrecht: Reidel) pp. 177-180.

HARRIS, D. L.   1961.   Photometry and colorimetry of planets and
     satellites.   In *Planets and Satellites,* eds. G. P. Kuiper
     and B. M. Middlehurst (Chicago: Univ. of Chicago Press)
     pp. 272–342.
HARTMANN, W. K., CRUIKSHANK, D. P., DEGEWIJ, J., and CAPPS, R. W.
     1981.   Surface materials on unusual planetary object Chiron.
     Icarus, submitted.
JOHNSON, P. E., GREENE, T. F., and SHORTHILL, R. W.   1978.   Narrow-
     band spectrophotometry of Ariel, Umbriel, Titania, Oberon, and
     Triton.   *Icarus* 36, 75–81.
KUIPER, G. P.   1949.   The fifth satellite of Uranus.   *Publ. Astron.
     Soc. Pacific* 61, 129.
MORRISON, D., and CRUIKSHANK, D. P.   1974.   Physical properties
     of the natural satellites.   *Space Sci. Rev.* 15, 641–739.
MORRISON, D., CRUIKSHANK, D. P., PILCHER, C. B., and RIEKE, G. H.
     1976.   Surface compositions of the satellites of Saturn from
     infrared photometry.   *Astrophys. J. Lett.* 207, L213–L216.
NICHOLSON, P. D., and JONES, T. J.   1980.   Two-micron spectro-
     photometry of Uranus and its rings.   *Icarus* 42, 54–67.
REITSEMA, H. J., SMITH, B. A., and WEISTROP, D. E.   1978.   Visual
     and near-infrared photometry of the Uranian satellites.   *Bull.
     Amer. Astron. Soc.* 10, 585 (Abstr.).
SINGER, S. F.   1975.   When and where were the satellites of Uranus
     formed?   *Icarus* 25, 484–488.
SINTON, W. M.   1972.   A near-infrared view of the Uranus system.
     Sky Telesc. 44, 304–305.
SINTON, W. M.   1977.   Uranus: The rings are black.   *Science* 198,
     503–504.
SMITH, B. A.   1977.   Uranus rings: An optical search.   *Nature*
     268, 32.
WHITAKER, E. A., and GREENBERG, R. J.   1973.   Eccentricity and
     inclination of Miranda's orbit.   *Mon. Not. R. Astron. Soc.*
     165, 15P–18P.

NOTE:

After the Bath conference and following completion of the above
text, some new results on the infrared spectra of Titania, Oberon,
and Umbriel were published by Soifer et al.[*] They used the Hale
5-m telescope to obtain spectra in the 1.5–2.5 μm region as presented here, and their results confirm the identification of water
ice or frost on the surfaces of the three satellites studied.   For
reasons described in their paper, they favor very low geometric
albedos for the Uranian satellites.

[*]SOIFER, B. T., NEUGEBAUER, G., AND MATTHEWS, K.   1981.   Near-
     infrared spectrophotometry of the satellites and rings of
     Uranus.   *Icarus* 45, 612–617.

THE RINGS OF URANUS

André BRAHIC, Observatoire de Paris, Université Paris VII

One of the very exciting discoveries in astronomy and planetary sciences in recent times is the detection of a series of narrow rings around Uranus. During more than three centuries, the rings of Saturn have had a special fascination and symbolism and an enormous amount of literature has been devoted to studies of their nature, properties and origin. The discovery of Uranus's rings, and two years later of Jupiter's rings has not only renewed interest but also raised a number of new cosmogonical questions.

Planetary rings are important not only because of the dynamical problems which they pose, but also because it is probable that processes which played a role in planet and satellite formation are still at work in these rings: rings systems are probably examples of arrested growth and they afford a good opportunity of studying some of the accretion mechanisms which operated in the early solar system. Planetary rings may also provide an appropriate analog for events in flat systems like spiral galaxies or accretion discs. Furthermore, particles of planetary rings are natural probes of the internal structure of the central planet.

Nobody knows if Uranian rings have been formed with Uranus $4.5 \ 10^9$ years ago or are younger. Any model attempting to understand the origin of these rings must be considered as very uncertain. We do not discuss here the problem of the origin of planetary rings (break-up of a large body, condensation of the protoplanetary nebula, meteoroidal bombardment, or a combination of these processes). A number of problems like the interaction with interplanetary micrometeoroids (their amount is not well known at the distance of Uranus) or the interaction with Uranus magnetosphere (which is not known - Brown, 1976) are not considered here. In this book, Elliot has reviewed the data and analyses of observations. For earlier reviews, see Ip (1980), Dermott (1981b), and Brahic (1981). In this brief article, deductions from observations are first surveyed. Then, we confine ourselves mainly to a discussion of the dynamics of narrow rings and on ring-satellite interactions. Finally a brief comparison of Uranian rings with Jovian and Saturnian rings is reviewed and a number of unsolved problems and future observations are quoted.

It is interesting to note that Herschell thought in 1787 and 1789 that he saw two stubby rings around Uranus as shown in Figure 1.

Further observations convinced him by 1798 that Uranus has no ring in the least resembling those of Saturn.

Figure 1. Herschell's drawings.

Many papers have been published just after the discovery of Uranian rings. I have no room here to quote all of them. Now, the dust has settled and I outline here only that which I consider as the most important results on ring dynamics, but I do not enter into any details of the models, which are described in the references at the end. I have only tried to emphasize the main physical arguments without any detailed calculation.

DISCOVERY OF THE RINGS OF URANUS

It is almost impossible to observe directly from the Earth a system of dark rings within 8 seconds of arc around a planet which has an apparent diameter of 4 seconds of arc. The Uranian rings have been discovered during the occultation by Uranus of the late type star SAO 158687 of visual magnitude -9.5 on March 10, 1977. High-speed photoelectric photometry of occultations provides a powerful tool for probing the upper atmosphere of a planet, as has been shown during the past decade from occultations involving Mars, Jupiter, and Neptune. When the occultation of a reasonably bright star by Uranus was predicted, teams of observers from United States, France, India, China, Japan, Australia, South Africa planned to travel to Africa, South Asia, and Australia, where the event would be visible. In addition, arrangements were made to fly N.A.S.A. Kuiper Airborne

Observatory with its 91-cm telescope to the southern hemisphere. All
of these efforts paid off when not only the predicted occultation by
the planet, but also a series of secondary occultations by the previously unsuspected rings, were observed. The more complete observations have been made at that time by Elliot and his team observing
from the N.A.S.A. plane over the southern Indian Ocean. It is interesting to note that the recorded conversation (Elliot et al., 1977a)
among the observers shows that they progressively changed from skep-
ticism to unbounded enthusiasm. Since, seven useful occultations
have been observed and the structure of the rings is now known with
an accuracy of the order of a few kilometers and less. It is pleasant to note that occultations techniques are very powerful: they
give a resolution slightly better than the one obtained from a
Voyager type spacecraft flying-by the planet. From the Earth, the
resolution is limited by Fresnel diffraction and by the angular size
of occulted star at the distance of Uranus. The "best" stars tend to
be giant stars, giving us a resolution limit of a few kilometers.
A more detailed review of the observations is given in Elliot's
paper in this volume. Here I concentrate only on the main deductions
from observations.

DEDUCTIONS FROM OBSERVATIONS

At least nine narrow rings encircle the planet, extending
between 1.60 and 1.95 planetary radii and identified in order of
increasing planetary distance (6,5,4,α, β, η, γ, δ, ϵ). Compared to
their circumference (some 250 000 km), they are exceedingly narrow:
most do not exceed 10 kilometers in width and only one, the outermost ring, spans as much as 100 km. Three rings are circular, but an
unexpected revelation is that at least six have eccentricities of
0.001 to 0.01 and have variable widths. Both of these characteristics are best illustrated by the external ring, called ϵ. It is the
largest and the best known: its distance from Uranus varies by about
800 km and its width changes from 20 to 100 km linearly with its distance from Uranus. This structure can be simply explained if ring
boundaries are, as shown in Figure 3, Keplerian ellipses with same
apsidal line, the eccentricity of the outer ellipse being a little

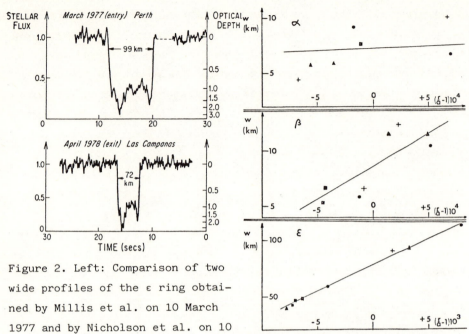

Figure 2. Left: Comparison of two wide profiles of the ε ring obtained by Millis et al. on 10 March 1977 and by Nicholson et al. on 10 April 1978. (After Nicholson et al., 1978). Right: Radial width of the rings α, β and ε plotted against radius for 7 occultations. (After Sicardy et al., 1981).

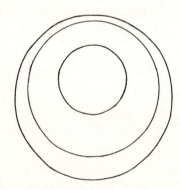

Figure 3. Shape of ε ring.

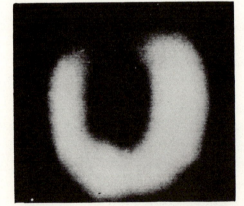

Figure 4. First picture of Uranus rings. (Nicholson et al., 1978).

bigger than the one of the inner ellipse.

The remarkable thing is that these elliptic rings precess slowly about the planet, just as they should due to the planet's oblateness. For example, the ε ring precesses at a rate of 1.364°

per day. Normally, the rate of precession would depend on the distance to the planet: the inner edge of a ring of any width would
precess more rapidly than its outer boundary. This differential precession would shear each ring into a circular band in a time scale
of only a few tens of years. In fact, each ring precesses as a rigid
whole, the eccentric nature of the rings seems to be continually re-
inforced, either by satellites or by the ring material itself. The
profile of the rings looks the same everywhere (Figure 2). The rings
have sharp outer edges and structure bigger than noise appears
within a number of rings.

The rings are really dark: they reflect less than a few percent of the light that strikes them: their visual albedo is of the
order of 0.025. This low albedo implies that the majority of ring
particles lack coatings of water, ammonia or methane frosts, materials which are common among many of the outer-planet satellites, including the Uranian satellites.

Near 2.2 microns, the methane absorption is enormous. It seems
there are almost no clouds in the atmosphere of Uranus and either
ammonia frosts or certain hydrated silicate minerals could cause
these well-defined absorption features. At 2.2 microns, the albedo
of Uranus is only 10^{-4} and the rings are brighter than the planet.
Occultations must be observed at this wavelength.

Because of their proximity to Uranus, small cross-section, and
low albedo, the rings are extremely difficult to see. In fact, using
the 5-meters telescope at Palomar mountain, Matthews, Nicholson and
Neugebauer (1981) have obtained in 1978 the first picture of the
ring system by substracting sets of scans made at 1.6 microns (where
the planet is brighter than the rings) from corresponding scans at
2.2 microns (where the planet absorbs virtually all the light that
strikes it). The ϵ ring has the most important contribution to this
image. Its signature can be seen clearly: the bottom of the image
corresponds to its wide part and the top to the narrow part (Figure
4). It is not yet understood why the image is slightly brighter on
the right. It seems that there is less material between the rings
than in the rings. An even smaller amount of material could easily dominate if it was there. The fact that we see the signature of the

rings indicates that spaces between the rings are relatively free of debris.

If we plot the precession of the α, β and ε rings against semi-major axis, we get a result just as it should be if it was a precession around an oblate planet. Any satellite able to affect this precession rate should be seen from the Earth. A value of 0.0034 for J_2 (Nicholson et al. 1978). From J_2 and the optical oblateness (difficult to observe from the Earth), the period of rotation of the planet can be deduced if hydrostatic equilibrium holds. A rotation period of 15.5h has been obtained by Elliot et al. (1981). Elliot et al (1981) give a value of $(-2.9 \overset{+}{-} 1.3) \cdot 10^{-5}$ (see also Nicholson et al., 1981) for J_4. This value is consistent with several interior models. If the error bar could be reduced by an order of magnitude, it could become an useful discriminator for interior models.

Additional features like detailed structure inside some rings (similar to the Saturn F ring?) and broad components outside two or three rings (Elliot et al.,1981 and Sicardy et al.,1981) have also to be explained.

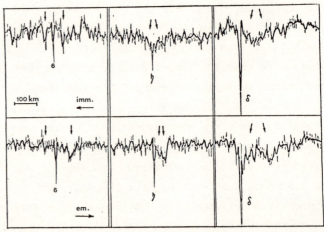

Figure 5. A broad component has been observed at the outer edge of η ring. Others are suspected close to 6 and δ rings. (After Sicardy et al., 1981).

Details on the data of the observed occultations can be found in the literature (Bhattacharyya & Bappu, 1977; Chen et al., 1978 Elliot et al., 1977,1978,1980,1981 ; Hubbard & Zellner, 1980; Mahra & Gupta, 1977; Millis et al., 1977,1978; Nicholson et al., 1978, 1981; Sicardy et al., 1981; Tomita, 1977).

RINGS DYNAMICS

Unconfined rings

It has been shown (Brahic, 1975,1977, 1981 and Goldreich and Tremaine,1978a,b) that,after a very fast flattening of the order of few tens of collisions per particle, a three-dimensional system of inelastically colliding particles reaches a quasi-equilibrium state in which the thickness of the newly formed disc is finite and in which collisions still occur. The combined effect of differential rotation and inelastic collisions makes the disc spread very slowly, particles moving both inwards and outwards carrying out some angular momentum. In the absence of external confining forces, the spreading time is of the order of the time it takes particles to random walk a distance equal to the ring width. The larger are the particles, the larger is the rate of this spreading (Brahic, 1977).

Poynting-Robertson drag (due to impacts with photons of light) and plasma drag (due to collisions with magnetospheric plasma) limit the lifetime of the small particles of a planetary ring (Burns et al 1981). But the rate of orbital decay caused by these processes is inversely proportional to particle size. Large particles are not affected by these drags.

If Uranian rings formed with the planet, the particles larger than a few millimeters would be eliminated by radial diffusion and the particles smaller than a few millimeters would have fallen onto the planet under the influence of Poynting-Robertson and plasma drags. Furthermore, differential shear and interparticle collisions should generate diffuse boundaries.

The presence of narrow rings with sharp edges around Uranus indicates that the rings are either young or confined. The absence of old diffuse rings seems to indicate that a confinement mechanism is present.

Resonances

A number of authors have considered resonances with known satellites. The figure 7, where the major satellite resonances are represented, is not very conclusive. There is not a simple one-to-

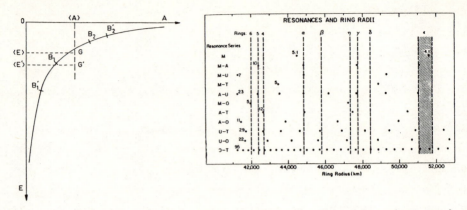

Figure 6. Angular momentum versus energy for a disc in which each particle is on a Keplerian orbit. The spreading disc is represented at the initial time by the arc B_1B_2 with barycentre G and at a latter time by the arc $B_1'B_2'$ with barycentre G'. This spreading is a direct consequence of the conservation of angular momentum, of the energy lost by collisions and of the concavity of the curve. (After Brahic, 1977).

Figure 7. Uranus's rings and satellites. Resonances and ring radii. (After Elliot et al., 1978).

one correspondence of resonant locations with ring structure. Furthermore the exchange of energy and angular momentum between the Uranian rings and the five known satellites of Uranus is not sufficient enough to confine the 9 narrow rings (Goldreich and Nicholson, 1977 and Aksness, 1977). Researchers now agree that resonances with moons outside the 9 rings are too weak to form most of the observed features directly.

Horseshoe orbits

Dermott, Gold et al. (1977,1979,1980) proposed the interesting idea that each narrow ring contains a small satellite which maintains solid particles in stable, horseshoe orbits (as seen by an

observer riding on the satellite) about its Lagrangian equilibrium
points (Figure 8). This study is probably very appealing for a
better understanding of the motion of the recently discovered co-
orbital satellites around Saturn. But inelastic collisions should
destroy a ring of colliding particles (Aksness, 1977 and Goldreich
and Nicholson, 1977) and it seems difficult to have a ring of confined particles at a location of a maximum of energy.

Cook and MacIntosh (1981) have proposed that satellites disintegrated inside Roche limit and are now dismembered into ring particles. The resulting condensed rings being a stable entity held
together by its own gravitation.

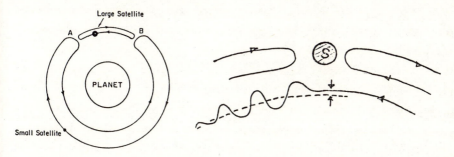

Figure 8. Left: Horseshoe orbit. (After Dermott, 1981).

Right: Ring particles paths in a frame rotating with the
perturbing satellite. (From Goldreich and Tremaine , 1981).

Guardian satellites

A very interesting mechanism which is probably the right explanation has been quantitatively studied by Goldreich and Tremaine
(1979,1980,1981) and is now explored by Hénon (1981), Brahic (1981),
Dermott (1981), Borderies (1981) and others.

Through gravity, a nearby satellite can alter a ring particle'
s orbit, making it elliptical. The overall effect of this on a popu-
lous disc of particles is to increase in some places the density of
particles and to decrease it in other places and finally to give
rise to spiral density waves, like those in galaxies. If the perturbing satellite is exterior to the ring, this wave moves   outwards,

carrying negative energy and angular momentum. An isolated ring particle would return to its original position, but the rings are dense enough to make collisions between particles, and particles involved in such collisions finally move inward toward the planet. The outside satellite tends to take angular momentum from the ring and put to its orbit. Conversely, an inner satellite, moving faster than the particles of the ring, adds energy to nearby particles and transfers angular momentum to the ring from its orbit. This exchange of angular momentum have been first studied by Lin and Papaloizou (1979). A satellite near a planetary ring can push the ring away and simultaneously the satellite is repulsed. Small undetected satellites on each side of a ring could constrain its edges and prevent ring spreading, elliptical rings can also be generated by such a confining mechanism.

It is not intuitively obvious how a force of attraction can act to repel another body. This repulsion mechanism between a disc and a satellite depends on the presence of many ring particles that can often collide with each other. Changing a particle's angular momentum and thus its orbit depends on its having collisions or gravitational interactions (close encounters) with other particles. The physics involving many colliding particles sets the study of ring dynamics apart from classic celestial mechanics, in which few or no collisions are assumed, and fluid dynamics, in which much more frequent collisions are known to occur than in a planetary ring (the mean free path of an individual particle is much shorter in a fluid).

Satellites exert torques on the ring material at the locations of their low order resonance. For the simplest case of a circular ring and a circular satellite orbit, the strongest torques occur where the ratio of the satellite orbit period to the ring particle period equals $\frac{m}{m + 1}$ where m is an integer. The torque is of the order of: $T_m \sim \pm m^2 \dfrac{G^2 m_s^2 \sigma}{\Omega^2 r^2}$ (Goldreich and Tremaine 1980,1981). G, Ω, m_s, r and σ are respectively the gravitational constant, the orbital frequency, the satellite mass, the radius and the surface mass density evaluated at the resonance location.

The spacing between neighbouring resonances from a nearby

satellite is very small. The widths of individual resonances are greater than the frequency separations so that there are resonances that overlap. Thus it is useful to sum the discrete resonances torques and to define the total torque on a narrow ringlet of width r:

$$T \sim \pm \frac{G^2 m_s^2 \sigma r \Delta r}{\Omega^2 x^4}$$

where x is the separation between the satellite and the ringlet (r >> x >> Δr) (Goldreich and Tremaine, 1980).

The presence of dissipation is necessary. Although the expression for the torque does not reveal explicitly any dependence on dissipation, T_m would vanish without dissipation. The exact nature of the dissipation does not affect the expression of the torque. Particles collisions are evidently the main source of dissipation.

This resonance torque does not act on isolated test particles. For example, there are not enough collisions in the asteroid belt and thus such a mechanism is not responsible for the formation of the Kirkwood gaps. The rate of transfer of angular momentum can also be calculated by a perturbative approach without reference to individual resonances. The gravitational force on a ring particle due to a satellite is only effective at close encounters. Since the tangential component of the relative velocity of the particle with respect to the satellite is reduced during encounter (Figure 8), angular momentum is exchanged with the net result that the ring experiences a torque (Lin and Papaloizou, 1979). During encounter, a particle initially moving on a circular orbit, acquires a radial velocity (Figure 8) and thereafter moves on a Keplerian ellipse. In a frame corotating with the perturbing satellite, all particles initially moving in circular orbits must follow similar paths after encounter. Thus each perturbing satellite generates a standing wave. In the inertial frame, each particle moves on an independent Keplerian ellipse, but the pericentres of these elliptical orbits and the phases of the particles on the orbits are such that the locus of the particles is a sinusoidal wave which moves through the ring with the angular velocity of the perturbing satellite. The damping of these waves, by collisions, results in a net exchange of angular momentum between the satellite and the ring particles.

This phenomenon is a sort of <u>viscous</u> <u>dynamical</u> <u>friction</u> studied in stellar dynamics. There are three main resonances for a Keplerian disc of infinite extent: one inner and one outer Lindblad resonance where the epicyclic motion of a particle in a circular orbit is strongly excited, since the perturbation frequency felt by the particle is equal to its epicyclic frequency and one corotation resonance where the angular momentum of one particle in a circular orbit undergoes large changes, since the particle feels a slowly varying azimuthal force (Figure 9).

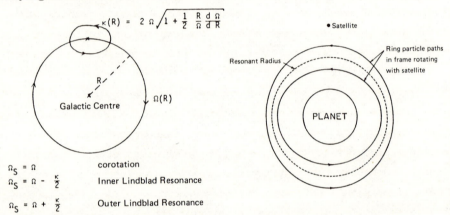

$$\kappa(R) = 2\,\Omega\,\sqrt{1 + \frac{1}{2}\,\frac{R}{\Omega}\,\frac{d\,\Omega}{d\,R}}$$

Galactic Centre $\Omega(R)$

$\Omega_S = \Omega$ corotation

$\Omega_S = \Omega - \dfrac{\kappa}{2}$ Inner Lindblad Resonance

$\Omega_S = \Omega + \dfrac{\kappa}{2}$ Outer Lindblad Resonance

• Satellite

Resonant Radius

Ring particle paths in frame rotating with satellite

PLANET

Figure 9. Left: Corotation and Lindblad resonances for galactic disc.
Right: Particle paths in a frame rotating with the perturbing satellite in the vicinity of a resonance. The density of particles increases on the left and on the right and decreases in perpendicular directions. (After Dermott, 1981).

Planetary rings can support leading and trailing spiral density waves which are controlled by a combination of the Coriolis force and of the ring's self gravity. This phenomenon is similar to density waves in Messier 51. Close to a resonance, the long spiral waves have wavelengths several orders of magnitude greater than the interparticle spacing. These waves can exist only on the satellite side of the resonance and propagate toward and away from the resonance. The satellite excites the long trailing wave at the resonance and

this wave carries away all of the angular momentum (positive or
negative) which the resonance torque gives to the disc. The wave
damps due to non linear and viscous effects close to the resonance.
The particles on the satellite side of the resonance move toward the
resonance. If the resonance torque is sufficiently large, a gap
opens on the satellite side of the resonance. Goldreich and Tremaine
(1978) have proposed this mechanism for producing the Cassini
division in Saturn's rings due to the influence of Mimas which
orbits well outside the rings.

Thus the structure of Uranian rings appears to require the
existence of small, as yet undiscovered satellites which orbit
within the ring system. We have no room here to describe in details
the quantitative study of the interactions of the rings and these
satellites, I give here only the main results of the brilliant work
of Goldreich and Tremaine (1979,1980,1981).

Torques not only transfer angular momentum but also energy,
they can also change the eccentricity of the rings. Then torques can
pump up the eccentricity of the rings, and the eccentricity of the
satellites increase also, in other cases torques can damp the eccentricities of both the rings and the satellite. The ε ring is probably associated with an eccentric satellite or two. If resonance torques are non linearly limited, both eccentricities of the ring and
of the satellite will be non zero. Near corotation resonance, the
eccentricity decreases. Near Lindblad resonance, the eccentricity
increases (Goldreich and Tremaine, 1981).

Four different mechanisms have been explored by Goldreich and
Tremaine (1979) to understand the precession of ε ring:

- The self gravity of the rings can maintain apse alignment if
the mass of the satellite is at least of the order of $5 \; 10^{18}$ g.
(that is much greater than that implied by the confinement theory)
and if the mean surface density at quadrature is of the order of 25
g. cm^{-2}.

- A nearly guardian satellite could force uniform precession
in the ring. Goldreich and Tremaine have calculated the parameters of
such a satellite.

- Smooth pressure gradients, due to interparticle collisions cannot produce uniform precession without an unreasonably large ring thickness.

- In a system of colliding particles, the possibility of a discontinuity analog to a shock which cannot be modelled by smooth pressure gradients shock-like phenomena could force uniform precession, but have not been analyzed.

Goldreich and Tremaine (1979) claim that apse alignment is maintained by the self-gravity of the ring. Their model predicts a strong variation in surface density profile as a function of true anomaly as a consequence of the non-uniform eccentricity gradient accross the ring. This prediction can be tested by future occultation observations.

In order to confine the observed narrow rings around Uranus, small satellite are sufficient. For example, kilometer-sized bodies are large and massive enough to confine the observed rings. They evidently cannot be observed from the Earth. If each Uranian ring is confined by at least two satellites, a minimum of 18 hypothetical satellites are necessary to confine all the rings. This has no observational support and it is probably why this theory of the guardian satellites has not been immediately accepted. In fact, the hypothetical satellites that confine the rings are not "external" satellites, but rather they are the largest "particles" in the rings.

Hénon (1981) has shown that, in a disc of particles of different sizes, a sufficiently massive particle will be able to push away the other disc particles, both towards the inside and the outside, and thus to clear completely an annular zone. As a consequence the particle population divides into:

a) the particles with a radius r larger than a critical radius (which is 2.6 km in the case of Saturn's rings, each of which describes an isolated orbit and has a clear gap;

b) a "continuous" distribution of particles with a radius r smaller than the critical radius and actually separated into a number of individual rings or ringlets by the massive particles. Brahic (1979,1981) is doing a numerical simulation of a system of colliding particles orbiting around a central mass and perturbed by a satellite. Contrary to other approaches like Monte-Carlo

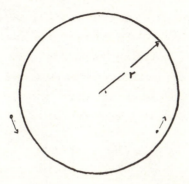

Figure 10. Two guardian satellites respectively inward and outward
a ring at a distance r from the center of the planet.

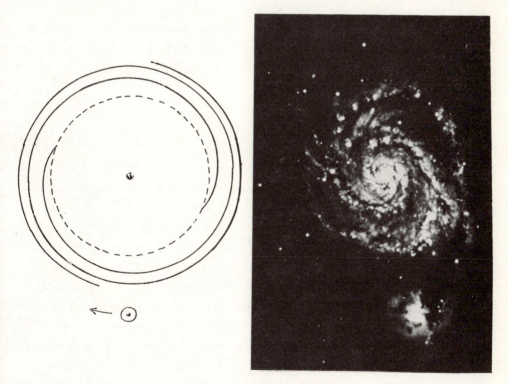

Figure 11. Left: a perturbing satellite generates a density wave
in a ring of colliding particles.

        Right: the galaxy Messier 51 and its companion which is
probably responsible for the grand design of waves.

methods, this model has the advantage to conserve very accurately the total angular momentum. The rate of transfer of angular momentum is calculated as a function of the mass of the satellite and its distance to the ring of colliding particles. The role of the most important resonances is particularly studied and the orbit of individual particles is calculated. There is no room here to enter into the details of the results which are presented elsewhere (Brahic, 1981).

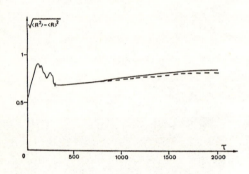

Figure 12. Spreading of the disc of colliding particles as a function of time. The full line represents the evolution of the disc without a perturbing satellite and the dashed line the evolution with a perturbing satellite (the disc is confined).

OTHER PLANETARY RINGS

The existence of ring systems around a planet is a natural consequence of collisions between ring particles and of the Roche limit. A cloud of particles moving around a central mass is rapidly flattened to a thin disc in a mean plane called the Laplacian plane which is identical to the equatorial plane of the planet in its vicinity. But unlike the planet's accumulation, large objects cannot grow within planetary rings because almost always these rings lie within what we call the Roche limit, inside which the self-gravity of an object is smaller than tidal stresses. Roche (1847) derived that, within a certain radial distance d to the centre of a planet of radius r and density u_p, a liquid satellite of density u_s in synchronous rotation around the planet. This distance is $2.446 \, (u_p / u_s)^{1/3} \, R$. The Roche limit, valid for an orbiting liquid body, has been used, often incorrectly, in the analysis of the break-up and accretion of solid bodies (see Smoluchowski, 1979 for a discussion). Because of their material strength, solid satellites can approach closer than the

Roche limit, but if they exceed a certain size, they are broken
apart by tidal forces (Aggarval and Oberbeck, 1974). The resulting
fragments will continually reimpact one another, so that the largest
member is not larger than some tens of kilometers.

We know now three systems of planetary rings. A quick look at
them can lead to the conclusion that they are not at all similar.
Seen from the Earth, Saturn has bright rings separated by narrow
gaps, like a negative print of the dark narrow Uranian rings separated by large gaps while the main Jovian ring, closer to its central
planet, is intermediate in size and has a three-dimensional halo.
The Neptunian ring is suspected but not yet discovered (and even
some authors claim that, as a consequence of the retrograde move-
ment of Triton, it does not exist). These three known systems have
been seen as perhaps more individualistic than the planet they
surround. In fact, Voyager exploration has revealed that Jovian and
Saturnian rings (Owen et al., 1979, Jewitt, 1981 and Science, 1981)
have many common characteristics with Uranian rings:

- Resonances: Like in the case of Uranian rings, most radial ring
structure of Jovian and Saturnian rings seems to be uncorrelated
with the simple perturbations of known satellites: some strongly
perturbed locations contain materials, other do not, some gaps are
at resonant positions, most are not. Only a few prominent features
of Saturn's rings appear to be associated with satellites'resonances.
The best example is given by the Cassini division, which contains
itself almost one hundred rings: Mimas clears a gap at the location
of the 2:1 resonance and there is a relative absence of matter in
the Cassini division, in comparison to adjacent parts of the A and B
rings (Goldreich and Tremaine, 1978).

- Narrow rings: Close-up photographs by Voyager disclosed Saturn's
rings to be more finely divided than anticipated. Rings lay within
rings: upwards a thousand have been counted but, there are probably
more at a scale narrower than Voyager could resolve. The previously
advanced explanation for the a priori perplexingly narrow rings of
Uranus can also explain the narrowest rings of Saturn and may
account for the grooved, "phonograph" appearance of the major rings'
subdivisions (Figure 13) (Goldreich and Tremaine, 1979,1980 and

Figure 13. The rings of Saturn in a computer-processed image   from Voyager 1. The outermost F ring can be seen on the left with one of the two guardian satellites that "bind" its edges.

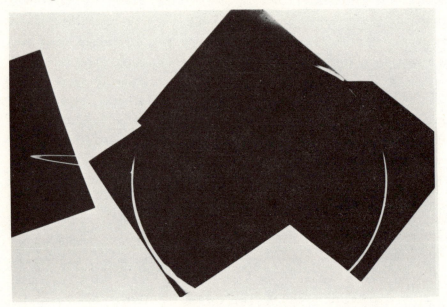

Figure 14. The ring of Jupiter seen by Voyager 2 from the planet's night side. The ring is unusually bright because of forward scattering from small particles within it. Similarly, the planet is outlined by sunlight scattered from a haze layer on the high atmosphere of Jupiter.

Figure 15. An elliptical ringlet (arrow) in the C ring of Saturn. The figure was made in joining together two Voyager 1 images of opposite ansae.

Hénon, 1981).

The Jovian ring is optically thin and appears to contain little structure. Because of smear motion, features smaller than 700 kilometres cannot be resolved in the Voyager images (Jewitt and Danielson, 1981).

Herding by satellites or larger than average ring particles (if the two can be distinguished any longer) seems to be a widespread phenomenon and may explain most of Saturn's and Uranus's ring structure. Large objects within the disc itself not only perturb nearby material and probably explain the intricate structure of the rings, but they can also serve as sources and sinks of ring matter (Burns et al., 1981).

- Eccentric rings: Voyager's images of Saturnian rings show one eccentric ring in the C ring of Saturn, another in the Cassini division, and perhaps a third in the F ring itself.

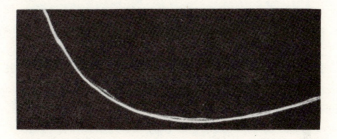

Figure 16. The F ring of Saturn. (N.A.S.A. document).

Figure 17. Segment of the F ring showing the position of the neighbouring satellites 1980 S 26 (left) and 1980 S 27 (right). The outer edge of the A ring is visible in the upper right. (N.A.S.A. document).

- Sharp edges: The discovery of small satellites just outside the Jovian ring lent credibility to theories which advocate the gravitational action of unseen satellites on the orbits of the rings' particles and this has been enhanced by the results of the Voyager 1 Saturn encounter. The outer A ring of Saturn and the main Jovian ring have both a sharp outer edge with a small satellite orbiting just at the outside of the ring, like a guardian satellite.

- Guardian satellites: The discovery of the narrow F ring around Saturn and of two nearby satellites 1980 S 26 and 1980 S 27 respectively outward and inward (Figure 17) is a great success for the guardian satellite theory. There is a 3500 kilometres gap between this narrow F ring (which Voyager resolved into three rings 30 to 50 kilometres wide, two of which appear to be interwoven) and the outer A ring. Although the actual geometrical relationships of these F rings and the "braided" structure are still to be worked out, it seems that the Uranian rings are not so unruly. Gravitational perturbations of the F ring particles' orbits by 1980 S 26 and 1980 S 27 not only account for confinement of the ring particles but probably explain some of the structure of the F ring. Goldreich (1981) notes that the herding satellites, being vastly larger than those suggested for Uranus, should set up waves in the rings, which could explain at least some of their irregular shapes. Furthermore microscopic particles of the F ring, whose presence is evidenced by the ring's brightness when back-lighted on Voyager's images, are electrically charged and pushed about by Saturn's magnetic field. We do not know yet the distribution of particle sizes in Uranian rings. Dermott (1981) has studied the wave pattern formed in a ring by a nearby satellite.

It is clear now that the same fundamental mechanisms (inter-particles' collisions, guardian satellites, resonances, electromagnetic forces on small particles,...) can explain much of the structure of the three known systems of rings. Nevertheless they are different and even inside Saturn the division into A, B and C rings is still meaningful and significant. The different distribution of sizes, the different nature of ring particles is not yet explained and is probably a result of different initial conditions and an illustration of how large is the number of solutions of Newton's equations.

ORIGIN OF THE SOLAR SYSTEM AND OTHER FLAT DISCS

Aside from satisfying our curiosity, studies of rings can shed some light on how the diffuse ball of gas and dust, from which the solar system formed, could have been subdivided and concentrated to produce planets, moons, asteroids, comets, and rings systems themselves.

We do not know if planetary rings are the result of the tidal break-up of a large satellite or of the halted accretion of primitive material. But, whatever are initial conditions, the dynamical processes occurring presently in planetary rings may provide an appropriate analog for events in the early solar system. Detailed structures visible today could be a fossil record of an intermediate stage in the accretion of orbiting bodies. Goldreich and Tremaine (1980) have presented an application of their study of disc-satellite interactions to the influence of the proto-Jupiter on the proto-planetary disc. In particular, the transfer of angular momentum in the protoplanetary disc and the role of resonances on the variations of orbits eccentricities (and thus on the relative velocities of the colliding planetoids) have to be explored in more detail. Models of planetary rings can be a good guide.

Techniques used for galactic dynamics have been particularly fruitful for the understanding of planetary rings. Conversely, a detailed study of rings structure can lead to a better understanding of spiral galaxies and accretion discs.

FUTURE WORK

Although the rings of Uranus may not be quite as mysterious as reported at first, much work remains for theoreticians who study ring dynamics. Goldreich and Tremaine made a particularly important step, but their study is only linear, we need now a non linear approach. It is difficult to predict future advances, but we can note that, so far, no calculations have been done in the case of large eccentricities. The transfer of angular momentum at the edge of a thick disc and the associated time scales have to be studied. The interaction of the magnetosphere of Uranus with small particles in

Uranian rings, the nature of the particles, the origin of the rings and many other problems have still to be solved.

New rings, new satellites, new features (detection of broad and diffuse components of the ring system) and much more detailed structures can be found in Uranian rings. It is necessary to observe many more occultations at least to have a coverage as complete as possible of the ring system. Klemola et al. (1981) and Mink et al. (1981) give the dates of the next star's occultations by Uranus and Neptune visible from the Earth. At this conference, Elliot discusses tha data we can expect from next occultations, Voyager and the Space Telescope.

In order to have a better understanding of planetary rings and more generally of processes which have led to the formation of planets and satellites, it should be particularly important to observe individual "stones" in the ring at least up to a few meters. Only a sophisticated orbiter can perform this. Galileo (for Jupiter 's rings) and S.O.P.2 (for Saturn's rings) will be already particularly useful.

The amount of work is still so large that we can again thank Herschell for its discovery, even if he told, a little incautiously in 1794: "...I therefore venture to affirm that Uranus has no ring in the least resembling that, or rather those, of Saturn."

I would like to express gratitude and thank very warmly many of my colleagues for all the information they gave me and for all the discussions we had. I acknowledge in particular J. Burns, A. Cazenave, J. Elliot, P. Goldreich, R. Greenberg, M. Hénon, J. Lecacheux, P. Nicholson, T. Owen, B. Sicardy, S. Tremaine.

It is not so easy to start a planetary ring with a crank-handle! (After Jean Eiffel).

REFERENCES

Aggarual,H.R., and Oberbeck,V.R. (1974). Roche limit of a solid body.
Astrophys.J. 191, 577-588.

Aksness,K. (1977). Quantitative analysis of the Dermott-Gold theory
for Uranus's rings. Nature 269, 783.

Bhattacharyya,J.C., and Bappu,M.K.V. (1977). Saturn-like ring system
around Uranus. Nature 270, 503-506.

Borderies,N. (1981), preprint.

Brahic,A. (1975). A numerical study of a gravitating system of colliding particles: applications to the dynamics of Saturn's
rings and to the formation of the solar system. Icarus 25,
452-458.

Brahic,A. (1977). Systems of colliding bodies in a gravitational
field. I. Numerical simulation of the standard model. Astron.
Astrophys; 54, 895-907.

Brahic,A. (1979). Influence of a satellite on the dynamical evolution of planetary rings. Bull. A.A.S. 11, 558.

Brahic,A., and Sicardy,B. (1981). Apparent thickness of Saturn's
rings. Nature 289, 447-450.

Brahic,A. (1981). Planetary rings.In Formation of planetary systems
(A. Brahic, Ed.)C.N.E.S. Paris.

Brahic,A. (1981). Preprint.

Brown,L.W. (1976). Astrophys. J. 207, L 209.

Chen,D.H., Yang,H.Y., Wu,Y.C., Kiang,S.Y., Huang,Y.W., Yeh,C.T.,
Chai,T.S., Hsieh,C.C., Cheng,C.S., and Chang,C. (1978). Photoelectric observation of the occultation of SAO 158687 by Uranian ring and the detection of Uranian ring signals from the
light curve. Scientia Sinica XXI, 503-508.

Cook,A.F., and Mac Intosh,B.A. (1981). The rings of Uranus. Nature,
submitted.

Dermott,S.F., and Gold,T. (1977). The rings of Uranus: theory.
Nature 267, 590-593.

Dermott,S.F., Gold,T., and Sinclair,A.T. (1979). The rings of Uranus
nature and origin. Astron. J. 84, 1225-1234.

Dermott,S.F., Murray,C.D., and Sinclair,A.T. (1980). The narrow
rings of Jupiter, Saturn and Uranus. Nature 284, 309-313.

Dermott,S.F., and Murray,C.D. (1980). Origin and eccentricity gradient and the apse alignment of the epsilon ring of Uranus. Icarus 43, 338-349.

Dermott,S.F. (1981a). The "braided" F ring of Saturn. Nature 290, 454-457.

Dermott,S.F. (1981b). The origin of planetary rings. Phil. Trans. R. In press.

Elliot,J.L., Dunham, E.W., and Mink,D.J. (1977a).The rings of Uranus Nature 267, 328-330.

Elliot,J.L., Dunham,E.W., and Millis,R.L. (1977b).Discovering the rings of Uranus. Sky Telesc. 53, 412-430.

Elliot,J.L., Dunham,E.W., Wasserman,L.H., Millis,T.L., and Churms,J. (1978). The radii of Uranian rings from their occultation of SAO 158687. Astron. J. 83, 980-992.

Elliot,J.L., Dunham,E., Mink,D.J., and Churms,J. (1980). The radius ellipticity of Uranus from its occultation of SAO 158687. Astrophys. J. 236,1026-1030.

Elliot,J.L., Frogel,J.A., Elias,J.H., Glass,I.S., French,R.G., Mink, D.J., and Liller,W. (1981). Orbits of nine Uranian rings. Astron. J. 86, 127-134.

Goldreich,P., and Nicholson,P.D. (1977). Revenge of tiny Miranda. Nature 269, 783-785.

Goldreich,P., and Tremaine,S. (1978). The velocity dispersion in Saturn's rings. Icarus 34, 227-239.

Goldreich,P., and Tremaine,S. (1978). The formation of Cassini division in Saturn's rings. Icarus 34, 240-253.

Goldreich,P , and Tremaine,S. (1979a).Towards a theory for the Uranian rings. Nature 277, 97-98.

Goldreich,P., and Tremaine,S. (1979b). Precession of the epsilon ring of Uranus. Astron. J. 243, 1638-1641.

Goldreich,P., and Tremaine,S. (1980). Disk-satellite interactions. Astrophys; J. 241, 425-441.

Goldreich,P., and Tremaine,S. (1981). The origin of the eccentricities of the rings of Uranus. Astrophys. J. 243, 1062-1075.

Hénon,M. (1981). preprint.

Hubbard,W.B., and Zellner,B.H. (1980). Results from the 10 March 1977 occultation by the Uranus system. Astron. J. 85, 1663-1669.

Ip,W.H. (1980). Physical studies of planetary rings. Space Sci. Rev. 26, 39-96.

Ip,W.H. (1980). New progress in the physical studies of planetary rings. Space Sci. Rev. 26, 97-109.

Jewitt, D.C. (1980). The rings of Jupiter. In the satellites of Jupiter (D. Morrison, Ed.). Univ. of Arizona Press, Tucson.

Jewitt,D.C., and Danielson,G.E. (1980). The Jovian ring. J. Geophys. Res., in press.

Klemola,A.R., Mink,D.J., and Elliot,J.L. (1981). Predicted occultations by Uranus: 1981-1984. Astron. J. 86, 138-140.

Lin,D.N.C., and Papaloizou,J. (1979). Tidal torques on accretion discs in binary systems with extreme mass ratios. Mon. Not. Roy. Astron. Soc. 186, 799-812.

Mahra,H.S., and Gupta,S.K. (1977). Occultation of SAO 158687 by Uranian rings. I.A.U. Circ. n° 3061.

Matthews,K., Nicholson,P.D., and Neugebauer,G. (1981). 2.2 microns maps of the rings of Uranus. preprint.

Millis,R.L., Wasserman,L.H., and Birch,P. (1977). Detection of the rings of Uranus. Nature 267, 330-331.

Millis,R.L., and Wasserman,L.H. (1978). The occultation of BD-15° 3969 by the rings of Uranus. Astron. J. 83, 993-998.

Mink,D.J., Klemola,A.R., and Elliot,J.L. (1981). Predicted occultations by Neptune: 1981-1984. Astron. J. 86, 135-137.

Nicholson, P.D., Persson,S.E., Matthews,K., Goldreich,P., and Neugebauer,G. (1978). The rings of Uranus: results of the 1978 10 April occultation. Astron. J. 83, 1240-1248.

Nicholson,P.D., Matthews,K., and Goldreich,P. (1981). The Uranus occultation of 10 June 1979. I. The rings. Astron. J. 86, 596.

Nicholson,P.D., Matthews,K., and Goldreich,P. (1981). Radial widths, optical depths and eccentricities of the Uranian rings,preprint

Owen,T., Danielson,G.E., Cook,A.F., Hansen,C., Hall,V.L., and Duxbury,T.C. (1979). Jupiter's ring. Nature 281, 442-446.

Science, vol.212, n° 4491, 10 April 1981.

Sicardy,B., Boucher,P., and Perrier, C. (1981). 15 August 1980 oc-
    cultation by Uranus. preprint.
Smoluchowski,R. (1979). Planetary rings. Comments Astrophys. 8,69-78
Tomita,K. (1977). Observation of occultation of the SAO 158687 star
    by Uranus at Dodaira station. Tokyo Astron. Bull. 250.

RINGS OF URANUS: A REVIEW OF OCCULTATION RESULTS
James L. Elliot, Massachusetts Institute of Technology

Since their discovery in 1977 (Elliot 1979), the dark, narrow
rings of Uranus have intrigued dynamicists. The main enigma has been
how the rings can remain so narrow - only a few km wide - when particle collisions and the Poynting-Robertson effect should cause the
particles to disperse. The Uranian rings have posed other problems
as well, and have proved to be a unique system for developing
dynamical models of rings. The reason for this theoretical interest
is the high precision and time coverage of the data available from
occultation observations. With occultations we obtain a spatial
resolution of 1 km in the position of ring segments and a resolution
of 4 km in their structural details. These high-resolution data are
available sufficiently often to be useful for dynamical purposes - at
the rate of 1-2 events per year. This spatial resolution is somewhat
better than that obtained by Voyager imaging of Jupiter's and
Saturn's rings (Owen et al. 1979; Smith et al. 1981). Ground-based
images of the Uranian rings, obtained by Matthews, Nicholson, and
Neugebauer (1981), have a spatial resolution of ~50,000 km. Although
unable to resolve individual rings, these data have established the
mean geometric albedo of the rings at 0.030 ± 0.005.

At this conference Brahic has reviewed the relation of the
Uranus rings to general models for rings and the ring systems of
Jupiter and Saturn. For an earlier review, see Ip (1980a, b).

In this chapter we review the structure and orbits of the nine
confirmed Uranian rings, as revealed by occultation observations from
10 March 1977 through 26 April 1981. We then compare the conclusions
from these observations with current dynamical models. Finally we
discuss what new information about the rings that we expect from the
following observing opportunities: (i) future occultations,
(ii) the Voyager encounter in 1986, and (iii) the Space Telescope.

OBSERVATIONS

The main method for observing the Uranian rings has been stellar occultations (Elliot 1979). With this technique we measure the intensity of a star versus time during the passage of the ring system through our line of sight to the star. The result of this measurement is the optical depth versus distance along a particular track through the ring system. The precision of the measured optical depth depends on the signal-to-noise ratio of the photometry; the spatial resolution of the optical depth is limited by either Fresnel diffraction or the angular subtent of the occulted star. At the mean opposition distance of Uranus (18 au) and for the mean wavelength of the K band (2.2μ), Fresnel diffraction causes the minimum width of an occultation profile to be 4 km FWHM (full-width at half-maximum). This fundamental limit to the spatial resolution is equivalent to an angular resolution of 3×10^{-4} arcseconds. If the occulted star has an angular diameter comparable to or larger than this, then the spatial resolution of the ring occultation profiles becomes worse than 4 km. As an example, a KO star with an angular diameter of 3×10^{-4} arcseconds has V = +9 and K = +7. Some occulted stars are large enough so that their angular diameters can be determined from the ring occultation profiles (Millis et al. 1977).

The observation of a stellar occultation requires a prediction of when and where it will be visible. The first publicized prediction of an occultation by Uranus was made by Gordon Taylor (1973), who predicted the now famous occultation of SAO 158687 on 10 March 1977 when the Uranian rings were discovered (see Elliot 1979 for a review). To make this prediction, Taylor used the technique of comparing the Uranus ephemeris with star positions in the SAO catalog.

Since occultations of SAO stars by Uranus occur infrequently, just after the discovery it looked like years before we would be able to observe the Uranus rings again. This dismal forecast was radically changed by two developments: first, occultations of stars fainter than the SAO catalog limit were predicted by Klemola and Marsden (1977), who compared the ephemeris of Uranus for 1977-1980 with plates taken with the Lick double astrograph. Second, the Cal

TABLE I.   OBSERVATIONS OF OCCULTATIONS BY THE URANIAN RINGS

Date	Star	Magnitude V	Magnitude K	Observatory	References
10 Mar 1977	SAO 158687	8.8	-	KAO[1]	Elliot, Dunham, and Mink (1977)
				Perth	Millis, Wasserman and Birch (1977)
					Hubbard and Zellner (1980)
				Cape Town	Churms (1977)
				Kavalur	Bhattacharyya and Kuppuswami (1977)
				Naini Tal	Mahra and Gupta (1977)
				Peking	Chen et al. (1978)
23 Dec 1977	KM 2[2]	10.4	-	Cabezón	Millis and Wasserman (1978)
10 Apr 1978	KM 5	11.6	10.1	Cerro Las Campanas	Nicholson et al. (1978)
10 Jun 1979	KM 9	-	11.5	Cerro Las Campanas	Nicholson, Matthews and Goldreich (1981a)
20 Mar 1980	KM 11	13.1	10.5	Cerro Tololo	Elliot et al. (1981a)
				Sutherland	"
15 Aug 1980	KM 12	-	8.7	Cerro Tololo	Elliot et al. (1981b)
				Cerro Las Campanas	Nicholson, Matthews and Goldreich (1981b)
				ESO[3]	Bouchet, Perrier and Sicardy (1980)
26 Apr 1981	KME 13[4]	-	7.5	See text	Not yet published

[1]Kuiper Airborne Observatory   [2]KM 2 = BD-15°3969   [3]European Southern Observatory   [4]KME 13 = BD-19°4222

Tech group (Persson et al. 1978) realized that the occultations of
most of Klemola and Marsden's stars could be observed with good
signal-to-noise ratio in K wavelengths (2.2μ), where Uranus has a
deep methane absorption band in its spectrum. In fact, the rings are
brighter than Uranus at these wavelengths, with mean opposition K
magnitudes of 12.0 and 13.1 respectively (Matthews, Nicholson, and
Neugebauer 1981). The predictions of occultations by Uranus from
Lick astrograph plates have been extended through 1984 by Klemola,
Mink and Elliot (1981).

As a result of these predictions and infrared techniques, occultations have been observed regularly from the discovery in 1977
through the present. These observations are summarized in Table I.
The most recent occultation, on 26 April 1981, was successfully
observed from Siding Spring, Mount Stromlo, Kavalur and Naini Tal.
At the time of this writing, the data have not been published.

What do the observations of Table I reveal as the components of
the Uranian rings system? The only structures confirmed by all
groups are nine narrow rings: 6, 5, 4, α, β, η, γ, δ and ε. The
cumbersome notation arose, as is often the case in these matters, for
historical reasons: see Elliot, Dunham and Mink (1977); Millis,
Wasserman and Birch (1977); Elliot et al. (1978); and Nicholson et
al. (1978). The occultation profiles of the nine rings are shown
at low spatial resolution in Figure 1; the high-resolution structure
of these rings will be discussed in the next section.

Other possible components of the ring system have been reported
by several observers, but remain unconfirmed. The criteria for
confirmation of ring structures are either (i) corresponding dips in
signal can be identified in the data from two sites, or (ii) for a
single site, corresponding dips having nearly identical structure are
observed at equal distances from Uranus, when projected into its
equatorial plane. Some latitude in the latter criterion would be
acceptable to allow for elliptical rings. Reports of dips in signal
that remain unexplained, have been given by Churms (1977), Tomita
(1977), Bhattacharyya and Bappu (1977), Chen et al. (1978), Millis
and Wasserman (1978), Bouchet, Perrier and Sicardy (1980), and
Hubbard and Zellner (1980). These dips have not been confirmed as

Figure 1. Occultations by the rings of Uranus. The pre-immer-
sion and post-emersion occultations by the rings of Uranus
observed with the Kuiper Airborne Observatory (Elliot,
Dunham and Mink 1977) have been plotted on the common scale of
distance from the center of Uranus in the ring plane. Occultations corresponding to the nine confirmed rings are easily
seen. Other possible occultation events are visible as shallow
dips on the individual traces. Much (if not all) of the low
frequency variations in the light curves are due to a variable
amount of scattered moonlight on the telescope mirror. (After
Elliot 1979).

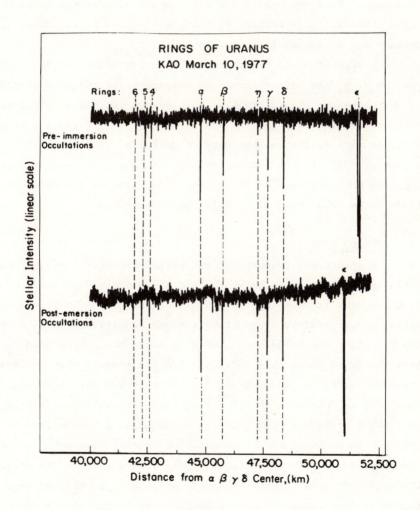

ring structures, and it is doubtful that they could have been caused
by small satellites, since none of the dips were deep enough to have
been a total occultation of the star. To maintain the small satellite hypothesis, one would have to assume that these dips in signal
were caused by grazing occultations or are due to diffraction
effects. Most dips lasted too long for the lack of total occultation
to be explained by diffraction, and it is highly unlikely that all
were caused by grazing occultations. Although the causes of these
dips in signal remain undetermined, mundane explanations are more
probable and have not been ruled out: small clouds, telescope
guiding errors, and starlight spilling out of the focal plane
aperture due to seeing fluctuations.

A system of broad rings around Uranus, with an optical depth as
great as 0.05, has been reported by Bhattacharyya and Bappu (1977)
and Bhattacharyya et al. (1980). This system remains unconfirmed by
other occultation data (Millis and Wasserman 1978; Nicholson et al.
1978), as well as the imaging data of Matthews, Nicholson and
Neugebauer (1981). The latter authors conclude that any material
composing such a system of broad rings must have an extremely low
albedo to be consistent with their image contours.

STRUCTURE

The structural features of the rings that have been the most
intriguing are their narrow widths and sharp edges. An example of
these features is shown by the profile of the γ ring in Figure 2
(Elliot et al. 1981b). The points are the occultation data, which
give the light transmission of the ring along the path probed by the
star; the solid curve is a model occultation profile that would have
been produced by an opaque ring, 3.4 kilometers wide. Although the
model ring is opaque, with a "rectangular" optical depth profile, its
occultation profile is not rectangular because of diffraction
effects. In particular, the occultation profile does not reach zero
intensity and diffraction fringes appear at the edge of the ring.
Diffraction fringes occur also at the edges of some of the other
rings - ε, for example - but the γ-ring profile has the largest
fringes, relative to its depth, of the five main rings. Hence, it

Figure 2. Occultation profile of the γ ring. The points represent 20-ms averages of the data and the solid curve is a
model profile for an occultation of a star 0.000 18 arcsec in
diameter by an opaque ring 3.4 km wide. The peaks at the
edges of the profile are due to Fresnel diffraction by the
abrupt boundaries of the ring. (After Elliot et al. 1981b).

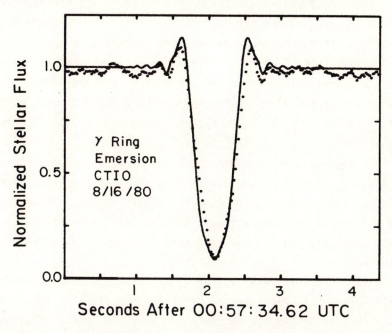

Figure 3. Occultation profile of the α ring. See text for
possible interpretations of the "double-dip" structure.
(After Elliot et al. 1981b).

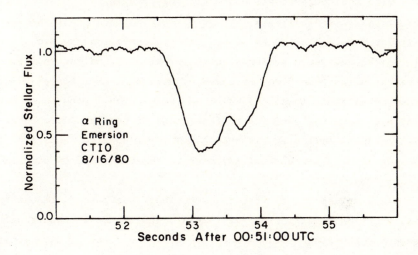

has more abrupt edges than the α, β, δ and ε rings. Precise
comparison with diffraction fringes produced by rings 6, 5, 4 and
the core of the η ring is difficult because of the generally lower
signal-to-noise ratio of these ring profiles.

Table II gives a summary of the observed ring structures.
Rings 6, 5, 4, γ, δ and the core of the η ring are narrower than the
resolution limit imposed by diffraction (4 km). The α ring sometimes
shows a "double-dip" profile, an example of which can be seen in
Figure 3. Resolution of the two peaks is barely above the
diffraction limit, so we do not know whether the α ring is a single
structured ring or two distinct rings, analogous to the "braided" F
ring of Saturn (Smith et al. 1981).

The ε ring is the broadest ring, whose width varies between 20
and 100 km. Nicholson et al. (1978) showed that the width correlated
linearly with the local radius of the ring, the broadest part occurring at apoapse and the narrowest part at periapse. This correlation
implies that the ε ring precesses as a unit, which we shall discuss
in the next section. The occultation profile of the ε ring shows an
undulating structure (Millis, Wasserman and Birch 1977) that has
remained the same over a period of years (Nicholson et al. 1978).
This structure is evident in the broad sections of the ring, but is
more difficult to discern in the narrow sections of the ring (for
reasons discussed by Elliot et al. 1981a).

The most bizarre structure yet observed is that of the η ring
(Figure 4). Originally, Elliot et al. (1978) reported it as a broad,
circular ring, about 50 km wide. Later observations by Nicholson et
al. (1978) indicated a narrow, circular ring of smaller radius. The
apparent disagreement was resolved by data of higher signal-to-noise
ratio obtained in August, 1980, which showed the η ring to be a broad
ring with a narrow core at the inner edge of the broad component
(Elliot 1979; Elliot et al. 1981b). The δ ring also shows a broad
section of diffuse material adjacent to the main, narrow ring (Elliot
et al. 1981b).

A simple model for the ring structure has been fitted to the
ring profiles by Elliot et al. (1981a), who assume a trapezoidal
shape for the ring profiles and include the effects of time constants

TABLE II.   SUMMARY OF RING STRUCTURE

Ring	Mean Optical Depth	Mean Width (km)	Comments
6,5,4	$\gtrsim 0.2$	<4	These rings have the lowest integrated optical depth and are the most difficult to detect.
α	0.8	7	Sometimes shows a "double-dip" structure (See Fig. 3); "braided" ring?
β	0.5	8	Occultation profile has flat bottom.
η	0.05	60	A broad ring containing an unresolved narrow core with mean optical depth $\gtrsim 0.2$.
γ	0.5	<4	Occultation profile shows the largest diffraction fringes, indicating sharper edges than the other rings.
δ	0.5	<4	Probable diffuse material adjacent to the main part of the ring.
ε	1 - 2	60	Width varies between 20 and 100 km; undulating structure in occultation profile.

Figure 4.   Occultation profile of the η ring.   The narrow component is unresolved and lies at the inner edge of the broad component, which is about 60 km wide.   (After Elliot et al. 1981b).

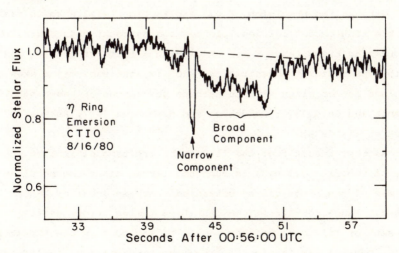

in the data recording and the angular diameter of the occulted star.
This model is adequate for obtaining ring widths and mid-occultation
times, but a more sophisticated model should include diffraction
effects and more realistic treatment of the α, ϵ and η rings.

RING ORBITS

Beginning with the original demonstration that narrow rings
encircle Uranus (Elliot, Dunham and Mink, 1977), a series of
improving kinematic models has been developed (Elliot et al. 1978;
Nicholson et al. 1978; Elliot et al. 1981a, b; Nicholson, Matthews
and Goldreich 1981a, b). The model of Elliot et al. (1981b) is fitted
by least squares to the midtimes of the ring occultations and has the
following free parameters: (i) the semi-major axis, a; orbital
eccentricity, e; and references longitude of periapse, ω_b; for each
ring; (ii) coefficients J_2 and J_4 that describe the zonal harmonics
of the Uranian gravity field; (iii) α_p and δ_p, the RA and Decl. of
the ring plane pole; (iv) corrections to the relative RA and Decl. of
Uranus and the occulted star for each event. Altogether 41
parameters were fitted to 105 data points, which span nearly a four
year interval.

The fitted parameters of interest are given in Table III, along
with their formal errors from the fit. The main conclusions of the
ring orbit solutions are (i) the rings fit the coplanar precession
model with an rms precision of 0.3 sec in time (or 2.8 km in radius);
(ii) the ring eccentricities, plotted against semi-major axis, show
a decreasing trend (except for the ϵ ring); (iii) rings γ, δ and η
possibly have "zero" eccentricity; and (iv) the semi-major axes of
the rings do not align with three body resonances of known satellites
(Dermott and Gold 1977; Goldreich and Nicholson 1977; Asknes 1977;
Elliot et al. 1978).

Another result from the fit of the orbit model is that the rms
error of the fit (0.3 sec) is much too large, since the occultation
midtimes of the rings can be determined with an accuracy better than
0.1 sec. Hence, either (i) unidentified timing errors exist in the
data set, or (ii) more effects should be included in the kinematic
model. Orbital inclinations are obvious parameters to include, but

TABLE III

FITTED MODEL PARAMETERS[1]

(a) Orbital Elements

Ring	Semi-Major Axis, a (km)	Eccentricity e $\times 10^3$	Azimuth of Periapse ω_o (degrees)[2]	Precession Rate From Fitted J_2 and J_4 (deg/day)	Precession Rate Fitted Individually (deg/day)
6	41863.8 ± 32.6	1.36 ± 0.07	235.9 ± 2.9	2.7600	2.7706 ± 0.0034
5	42270.3 ± 32.6	1.77 ± 0.06	181.8 ± 2.5	2.6678	2.6614 ± 0.0030
4	42598.3 ± 32.7	1.24 ± 0.09	120.1 ± 2.7	2.5963	2.5957 ± 0.0031
α	44750.5 ± 32.8	0.72 ± 0.03	331.4 ± 2.8	2.1832	2.1785 ± 0.0043
β	45693.8 ± 32.8	0.45 ± 0.03	231.3 ± 4.0	2.0288	2.0272 ± 0.0064
η	47207.1 ± 32.9	(0.03 ± 0.04)	(291.7 ± 88.3)	1.8094	. . .
γ	47655.4 ± 32.9	(0.04 ± 0.04)	(301.6 ± 49.1)	1.7503	. . .
δ	48332.0 ± 33.0	0.054 ± 0.035	139.0 ± 30.2	1.6657	. . .
ε	51179.7 ± 33.8	7.92 ± 0.04	215.6 ± 0.5	1.3625	1.3625 ± 0.0004

[1] For $M_u = 8.669 \times 10^{28}$ gm and $G = 6.670 \times 10^{-8}$ dyn cm^2 gm^{-2}

[2] At 20:00 UT on 10 March 1977

(b) Harmonic Coefficients of the Gravity Potential[3]

$$J_2 = (3.352 \pm 0.006) \times 10^{-3}$$

$$J_4 = (-2.9 \pm 1.3) \times 10^{-5}$$

(c) Pole of the Ring Plane

$$\alpha_{1950.0} = 5^h \ 06^m \ 26.1^s \pm 10.7^s$$

$$\delta_{1950.0} = +15° \ 13' \ 15" \pm 3'3"$$

[3] For a reference radius, R = 26,2000 km

This table was adapted from Elliot et al. (1981b).

are probably not the main effect causing the large residuals.

The final topic we discuss for the ring orbits is uniform precession, alluded to earlier in reference to the ε ring. Since the precession rate of a ring is a decreasing function of semi-major axis (see Eq. 1 of Elliot et al. 1981b), a ring of sufficient ellipticity would be disrupted by the different precession rates of the inner and outer particle orbits - unless the same precession rate for the entire ring is maintained by a force between ring particles in different orbits. By demonstrating the linear relation between width and radius for the ε ring, Nicholson et al. (1978) showed that ε ring precesses uniformly. The α and β rings have also been shown to be in uniform precession (Elliot et al. 1981a).

RESULTS FOR URANUS

From occultation observations, we obtain the following information about Uranus: (i) the coefficients J_2 and J_4 of the Uranian gravity potential; (ii) the coordinates (α_p, δ_p) of the ring plane pole; (iii) the equatorial radius, R_e, and ellipticity, ε; and (iv) the rotation period (from ε and J_2 under the assumption of hydrostatic equilibrium). The coefficients J_2 and J_4 are obtained directly from the ring orbit model (Table III) and can be used to constrain acceptable interior models for Uranus (MacFarlane and Hubbard, this conference; Podolak, this conference). Also obtained directly from the ring orbit model are the coordinates of the ring plane pole (Table III), which we can assume are also the coordinates of the Uranian north pole. For an unknown reason, these coordinates differ by several mean errors from the coordinates of the pole obtained from the satellite orbits (Dunham 1971, Elliot et al. 1981b).

The equatorial radius and ellipticity of Uranus are obtained from planetary occultation data, with the ring coordinate system as a reference. This allows chords from all occultations to be used in a single solution, which would not be possible without the rings. The values obtained by Elliot et al. (1981b), R_e = 26,145 ± 30 km and ε = 0.024 ± 0.003, refer to the occultation level (~8 x 10^{13} cm^{-3} for an atmosphere of hydrogen and helium in their solar proportions).

These values supercede the original result of Elliot et al. (1980)
and correct a systematic error in the earlier results. The occultation ellipticity agrees well with the value 0.022 ± 0.001 obtained
by Franklin et al. (1980) from a reanalysis of the Stratoscope II
images. The equatorial radius obtained from Stratoscope II by
Danielson, Tomasko and Savage (1972), 25,900 ± 300 km, refers to
the cloudtop level, which lies about 500 km below the occultation
level. Hence, the cloudtop level implied by the occultation result
is 25,650 km, which agrees with the Stratoscope II radius within its
error.

The rotation period determined from ε and J_2 under the assumption of hydrostatic equilibrium (see Eq. 3 of Elliot et al. 1980) is
15.5 ± 1.3 hours. By the same method, the period obtained from the
Stratoscope II ellipticity is 16.3 ± 0.5 hours (Franklin et
al. 1980). One may question whether Uranus is sufficiently close to
hydrostatic equilibrium for these estimates of the rotation period to
be reliable. As a test of the method, we can use Jupiter because J_2,
ε and its rotation period have each been measured independently.
Jupiter's rotation period is $9^h 55^m$. This can be compared to $9^h 47^m$,
the rotation period calculated from values of J_2 and ε given by
Smoluchowski (1976) for the 1 bar pressure level. For Saturn, the
period measured by Voyager 1 is $10^h 39^m$ (Desch and Kaiser 1981),
which can be compared to the value $10^h 29^m$ that we calculated from ε
and J_2 values measured by Pioneer (Gehrels et al. 1980, Anderson et
al. 1980, Null et al. 1981). Hence, the assumption of hydrostatic
equilibrium yields periods within 10^m of the correct ones for Jupiter
and Saturn, and we can be optimistic about the success of this method
for the rotation period of Uranus.

Periods of about 16 hours are obtained from some modern
spectroscopic results (Munch and Hippelein 1979; Brown and Goody
1980), while other modern spectroscopic results (Hayes and Belton
1977; Trafton 1977) and unpublished photometric data (Smith and
Slavsky 1979) favor longer periods. Our present knowledge of the
rotation period has been reviewed by Goody at this conference.

RELATIONS TO DYNAMICAL MODELS

Several structural features and orbital properties of the rings
have one or more proposed dynamical explanations. The sharp edges
and narrow widths are apparently explained by small satellites that
orbit nearby or within the rings. This idea has three possible
scenarios. First, the model of Goldreich and Tremaine (1979a)
postulates that each ring is constrained by two satellites, one with
an orbit inside the ring and the other with an orbit outside the
ring. Dermott, Gold and Sinclair (1979) have investigated a model
that postulates a single constraining satellite within each ring.
Since this model predicts an optical depth profile that is
symmetrical about the mean radius of a ring cross-section, the model
could not apply to the η ring (Figure 4). Finally, Cook and McIntosh
(1981) propose that each ring was formed from a single satellite that
is now dismembered into ring particles, so that each ring is held
together by its own gravity. The gaseous model of the rings (Van
Flandern 1979, 1980) is apparently untenable for reasons given by
Hunten (1980), Fanale et al. (1980) and Gradie (1980).

Another effect requiring of a dynamical explanation is the uni-
form precession of the ϵ, β and α rings. After examining several
mechanisms, Goldreich and Tremaine (1979b) conclude that self-gravity
is the most likely cause. Their model, combined with the orbital
parameters of the ϵ ring, implies a ring mass of 5 x 10^{18} gm. For
the β ring, Elliot et al. (1981a) obtained a mass of 4 x 10^{16} gm,
assuming the self-gravity model. The mass of the α ring is likely
similar to that of the β ring (Elliot et al. 1981a).

On the basis of their dual satellite model, Goldreich and
Tremaine (1981) have investigated what eccentricities would be
expected for the ring orbits. They find that each ring would have a
critical value of initial eccentricity, which would depend on the
masses and distances of the constraining satellites. If the initial
eccentricity is below the critical value, it would be damped to
zero - forming a circular ring. Rings η, γ and δ might be examples.
If the initial eccentricity is above the critical value, then it
would increase as time goes on. So far, no explanation has been
offered for the apparent decreasing trend (except for the ϵ ring) of

ring eccentricities with increasing semi-major axis (see Figure 7 of Elliot et al. 1981b).

The unusual structure of the η ring (Figure 4) has been discussed at this conference by Cook and Dermott, but no models have yet been published.

PROSPECTS FOR THE FUTURE

For the next few years, we can expect new observations of the Uranian rings from three sources: occultations, Voyager and the Space Telescope.

Probably the most significant result that we can anticipate from further Earth-based occultation observations, would be the undisputed detection of a broad, diffuse component of the ring system. Such material must exist, at some level, as can be seen from the following argument. Relative to gravity, the magnitude of non-gravitational forces - such as the Poynting-Robertson effect and particle collisions - become greater for smaller particles. Hence, for some particle size, the non-gravitational forces will exceed the gravitational forces binding the rings, and the smaller particles will "leak out" from a narrow ring. These escaped particles will form a diffuse component of the ring system. The precise photometry required to detect this diffuse component is occasionally possible from the ground and can be routinely achieved with the Kuiper Airborne Observatory. Other knowledge to be gained from further ground-based occultation observations would be improved precision of ring orbits and optical depth profiles. Also, we need to find the reason for the unaccountably large residuals from the ring orbit solution, which might be caused by a yet unidentified dynamical effect.

Voyager's flyby of the Uranus system in 1986 will yield a variety of unique information. First, it should be able to locate satellites within the ring system and thereby tell us which (if any) of the constraining satellite models is correct. Also, Voyager imaging should be more sensitive than occultations to broad rings of low optical depth. Voyager images can resolve individual rings and yield the reflectance of each ring for a variety of wavelengths over a large range of phase angles. Although the resolution of the ring

images will be only ~30 km due to smear, a stellar occultation
observed with the photopolarimeter could achieve a spatial resolution
of a few hundred meters (Stone, this conference). Because of
constraints on the trajectory required to continue the mission to
Neptune, values of J_2 and J_4 obtained by Voyager are expected to be
much less precise than those that have been obtained from the ring
precessions (see Table III and Stone, this conference).

The Space Telescope (ST), currently scheduled for launch in
1985 (Caldwell, this conference), will be a useful instrument for
observation of the rings because of its access to far uv wavelengths
and excellent resolution (about 0.1 arcsecond, which corresponds to
1000 km at the distance of Uranus). Hence, the ST should be able to
obtain a reflection spectrum of the ϵ ring and perhaps several other
rings as well. The resolution of the ST will also allow rejection of
the background light from Uranus so that occultations can be observed
in the far red and uv with much better signal-to-noise than can be
achieved from the ground. On the ST, the High Speed Photometer (HSP)
is being equipped with a red-sensitive photomultiplier (with a GaAs
photocathode), particularly to observe stellar occultations by outer
solar system bodies with methane absorption bands in their spectra.
The occultation data in the uv will have three times better spatial
resolution than the 2.2μ data because of the smaller size of Fresnel
diffraction effects. The occultation profiles of the rings at uv and
near ir wavelengths can be compared and will possibly yield
information about the particle sizes in the rings. The superior
resolution of the ST might also permit one of its cameras to detect
small satellites within the ring system.

The ever more revealing observations of the Uranian rings
through occultations, Voyager and the Space Telescope should continue
to inspire progress in dynamical models for narrow rings and to
improve our understanding of Uranus itself. And it is appropriate
that these opportunities should help us celebrate the two-hundredth
anniversary of Hershel's discovery.

I thank Dr. R.G. French for his suggestions. At M.I.T. Uranus
ring research is supported by grants from NASA and NSF.

REFERENCES

Asknes, K.   (1977).   Quantitative analysis of the Dermott-Gold
     theory for Uranus's rings.   Nature 269, 783.

Anderson, J.D., Null, G.W., Biller, E.D., Wong, S.K., Hubbard, W.B.
     and MacFarlane, J.J.   (1980).   Pioneer Saturn celestial mechan-
     ics experiment.   Science 207, 449-452.

Bhattacharyya, J.C. and Bappu, M.K.V.   (1977).   Saturn-like ring
     system around Uranus.   Nature 270, 503-506.

Bhattacharyya, J.C., Bappu, M.K.V., Mohin, S., Mahra, H.S. and Gupta,
     S.K.   (1979).   Extended ring system of Uranus.   Moon and
     Planets 21, 393-404.

Bouchet, P., Perrier, C. and Sicardy, B.   (1980).   Occultation by
     Uranus.   I.A.U. Circ. No. 3503.

Brown, R.A. and Goody, R.M.   (1980).   The rotation of Uranus.   II.
     Astrophys. J. 235, 1066-1070.

Chen D.-H., Yang H.-Y., Wu C.-H., Wu Y.-C., Kiang S.-Y., Huang Y.-W.,
     Yeh C.-T., Chai T.-S., Hsieh C.-C., Cheng C.-S. and Chang C.
     (1978).   Photoelectric observation of the occultation of
     SAO 158687 by Uranian ring and the detection of Uranian ring
     signals from the light curve.   Scientia Sinica XXI, 503-508.

Churms, J.   (1977).   Occultation of SAO 158687 by Uranian satellite
     belt.   I.A.U. Circ. No. 3051.

Cook, A.F. and McIntosh, B.A.   (1981).   The rings of Uranus.   Nature
     (submitted).

Danielson, R.E., Tomasko, M.G. and Savage, B.D.   (1972).   High resol-
     ution imagery of Uranus obtained from Stratoscope II.   Astro-
     phys J. 178, 887-900.

Dermott, S. and Gold T.   (1977).   The rings of Uranus: theory.
     Nature 267, 590-593.

Dermott, S.F., Gold, T. and Sinclair, A.T.   (1979).   The rings of
     Uranus: nature and origin.   Astron. J. 84, 1225-1234.

Desch, M.D. and Kaiser, M.L.   (1981).   Voyager measurement of the
     rotation period of Saturn's magnetic field.   Geophys. Res.
     Lett. (submitted).

Dunham, D.W.   (1971).   Motions of the satellites of Uranus.   Ph.D.
     thesis, Yale University.

Elliot, J.L.   (1979).   Stellar occultation studies of the solar
     system.   Ann. Rev. Astron. & Astrophys. 17, 445-475.

Elliot, J.L., Dunham. E.W. and Mink D.J.   (1977).   The rings of
    Uranus.   Nature 267, 328-330.

Elliot, J.L., Dunham , E.W., Wasserman, L.H., Millis, R.L. and Churms,
    J.   (1978).   The radii of Uranian rings α, β, γ, δ, ϵ, η, 4, 5
    and 6 from their occultation of SAO 158687.   Astron. J. 83,
    980-992.

Elliot, J.L., Dunham, E., Mink, D.J. and Churms, J.   (1980).   The
    radius and ellipticity of Uranus from its occultation of SAO
    158687.   Astrophys. J. 236, 1026-1030.

Elliot, J.L., Frogel, J.A., Elias, J.H., Glass, I.S., French, R.G.,
    Mink, D.J. and Liller, W.   (1981a).   The 20 March 1980 occul-
    tation by the Uranian rings.   Astron. J. 86, 127-134.

Elliot, J.L., French, R.G., Frogel, J.A., Elias, J.H., Mink, D.J.
    and Liller, W.   (1981b).   Orbits of nine Uranian rings.
    Astron. J. 86, 444-455.

Fanale, F.P., Veeder, G., Matson, D.L. and Johnson, T.V.   (1980).
    Rings of Uranus: proposed model is unworkable.   Science 208,
    626.

Franklin, F.A., Avis, C.C., Columbo, G. and Shapiro, I.I.   (1980).
    The geometric oblateness of Uranus.   Astrophys. J. 236, 1031-
    1034.

Gehrels, T. et al.   (1980).   Imaging photopolarmeter on Pioneer
    Saturn.   Science 207, 434-439.

Goldreich, P. and Nicholson, P.D.   (1977).   Revenge of tiny Miranda.
    Nature 269, 783-785.

Goldreich, P. and Tremaine, S.   (1979a).   Towards a theory for the
    Uranian rings.   Nature 277, 97-99.

Goldreich, P. and Tremaine, S.   (1979b).   Precession of the ϵ ring
    of Uranus.   Astron. J. 84, 1638-1641.

Goldreich, P. and Tremaine, S.   (1981).   The origin of the eccen-
    tricities of the rings of Uranus.   Astrophys. J. 243, 1062-
    1075.

Gradie, J.   (1980).   Rings of Uranus: proposed model is unworkable.
    Science 207, 626-627.

Hayes, S.H. and Belton, M.J.S.   (1977).   The rotational periods of
    Uranus and Neptune.   Icarus 32, 383-401.

Hubbard, W.B. and Zellner, B.H.   (1980).   Results from the 10 March
    1977 occultation by the Uranus system.   Astron. J. 85, 1663-
    1669.

Hunten, D.M.   (1980).   Rings of Uranus: proposed model is unwork-
     able.   Science 208, 625-626.

Ip, W.-H.   (1980a).   Physical studies of planetary rings.   Space Sci.
     Rev. 26, 39-96.

Ip, W.-H.   (1980b).   New progress in the physical studies of plane-
     tary rings.   Space Sci. Rev. 26, 97-109.

Klemola, A.R. and Marsden, B.G.   (1977).   Predicted occultations by
     the rings of Uranus 1977-1980.   Astron. J. 82, 849-851.

Klemola, A.R., Mink, D.J. and Elliot, J.L.   (1981).   Predicted
     occultations by Uranus: 1981-1984.   Astron. J. 86, 138-140.

Mahra, H.S. and Gupta, S.K.   (1977).   Occultation of SAO 158687 by
     Uranian rings.   I.A.U. Circ. No. 3061.

Matthews, K., Nicholson, P.D. and Neugebauer, G.   (1981).   2.2 Micron
     maps of the rings of Uranus.   (preprint).

Millis, R.L., Wasserman, L.H., Elliot, J.L. and Dunham, E.W.   (1977).
     The rings of Uranus: their widths and optical thicknesses.
     Bull. Amer. Astron. Soc. 9, 498.

Millis, R.L., Wasserman, L.H. and Birch, P.   (1977).   Detection of
     the rings of Uranus.   Nature 267, 330-331.

Millis, R.L. and Wasserman, L.H.   (1978).   The occultation of
     BD-15°3969 by the rings of Uranus.   Astron. J. 83, 993-998.

Münch, G. and Hippelein, H.   (1979).   The effects of seeing on the
     reflected spectrum of Uranus and Neptune.   Astron. & Astrophys.
     81, 189-197.

Nicholson, P.D., Matthews, K. and Goldreich, P.   (1981a).   The Uranus
     occultation of 10 June 1979.   I.   The rings.   Astron. J. 86,
     596-606.

Nicholson, P.D., Matthews, K. and Goldreich, P.   (1981b).   Radial
     widths, optical depths and eccentricities of the Uranian rings.
     Astron. J. (submitted).

Nicholson, P.D., Persson, S.E., Matthews, K., Goldreich, P. and
     Neugebauer, G.   (1978).   The rings of Uranus: results of the
     1978 10 April occultation.   Astron. J. 83, 1240-1248.

Null, G.W., Lau, E.L., Biller, E.D. and Anderson, J.D.   (1981).
     Saturn gravity results obtained from Saturn Pioneer II tracking
     and Earth-based Saturn satellite data.   Astron. J. 86, 456-468.

Owen, T., Danielson, G.E., Cook, A.F., Hansen, C., Hall, V. and
    Duxbury, T.C. (1979). Jupiter's rings. Nature 281, 442-446.

Persson, E., Nicholson, P., Matthews, K., Goldreich, P. and
    Neugebauer, G. (1978). Occultations by Uranian rings. I.A.U.
    Circ. No. 3215.

Smith, B.A. et al. (1981). Encounter with Saturn: Voyager imaging
    science results. Science 212, 163-190.

Smith, H.J. and Slavsky, D.B. (1979). Rotation period of Uranus.
    Bull. Amer. Astron. Soc. 11, 568.

Smoluchowski, R. (1976). Origin and structure of Jupiter and its
    satellites. In Jupiter (T. Gehrels, ed.), pp. 3-21.
    University of Arizona Press, Tucson.

Taylor, G.E. (1973). An occultation by Uranus. J. Brit. Astron.
    Assoc. 83, 352.

Tomita, K. (1977). Observation of occultation of the SAO 158687
    star by Uranus at Dodaira Station. Tokyo Astr. Bull., Sec.
    Series, No. 250.

Trafton, L. (1977). Uranus' rotational period. Icarus 32, 402-412.

Van Flandern, T.C. (1979). Rings of Uranus: invisible and impos-
    sible? Science 204, 1076-1077.

Van Flandern, T.C. (1980) Rings of Uranus: proposed model is
    unworkable. Science 208, 627.

FUTURE OBSERVATIONS OF URANUS

URANUS SCIENCE WITH SPACE TELESCOPE

John Caldwell*

Institute for Astronomy, University of Hawaii
2680 Woodlawn Drive, Honolulu, Hawaii 96822

ABSTRACT

The Space Telescope Observatory, scheduled for launch in 1985, is described. The advantages of the space environment and the consequent features of ST performance are given, with Uranus observations as examples. The first generation instruments, including two cameras, two spectrographs and a high speed photometer, are discussed. The Space Telescope Science Institute, which will manage the Observatory, is discussed briefly. The potential scientific interaction with the Voyager 2 encounter of Uranus is also considered.

INTRODUCTION

If the circum-Galactic orbit of our Sun had been slightly different in the astronomical past than it actually was, and if stellar perturbations had shorn our Solar System of all its members beyond ten A.U. from the center, then the Copernican model of the System would have been essentially complete as well as correct. If William Herschel had lived in a Solar System such as this, he would still have been a great astronomer. His quality is measured not merely by his discovery of Uranus, which was a singular and unplanned event in his career, but more significantly by his perception that previous improvement in telescope technology had produced major astronomical advances, and that continued efforts to build a better telescope offered the best prospect for further progress. Given the reality of the Solar System, his discovery

*Permanent address: E.S.S. Dept., S.U.N.Y. at Stony Brook, New York 11794.

of Uranus was an inevitable result of his approach to his science.

The subject of this paper is also an effort to build a better telescope. It is highly appropriate that a symposium honouring the work of William Herschel should include a contribution of this type. The Space Telescope Observatory (hereafter, ST), which includes a 2.4-meter primary mirror, will be launched into low Earth-orbit by the United States National Aeronautics and Space Administration Space Transportation System (Shuttle). The current schedule has the launch in the first quarter of 1985. The ST will include five scientific instruments, which are replaceable while in orbit. To be consistent with the astronomical potential of its space environment, the primary mirror will be by far the most precisely figured optical element of this size ever built. It is conservatively expected that the ST will produce revolutionary advances in understanding the Universe that are comparable to those due to any other single telescope ever constructed in the entire history of astronomy.

Space Telescope has been the subject of several recent papers. Spitzer (1979) has given a historical summary of the embryonic phase of the project and O'Dell (1981) has given a current overview. An entire volume (Longair and Warner, 1979) has been written concerning the scientific potential of the ST in all branches of astronomy. In that volume, O'Dell and Bahcall summarized the capabilities of the first generation scientific instruments. Among the review papers were ones by Belton on major planet astronomy and by Morrison on observations of smaller bodies. Leckrone (1980) has separately given a detailed account of the characteristics of the first generation instruments and Longair (1979) has described some of the scientific opportunities with ST. Finally, Caldwell (1981) has also summarized the observatory capabilities, using specific objectives for Uranus and Neptune as illustrative examples.

Because of the general availability of such papers, there has been no rigourous attempt here to give complete details of the ST, such as would be required, for example, to prepare an observing proposal. Rather, it is hoped that this paper will ignite the

interest of those who may not previously have been aware of the
unique capabilities of Space Telescope nor of its imminent arrival
at operational readiness.

ADVANTAGES OF THE SPACE ENVIRONMENT
Spatial Resolution

It has long been realized that local refractive inhomogeneities in the Earth's atmosphere distort images of celestial objects
far more severely than diffraction effects for all mirrors or
lenses larger than about 10 centimeters. However, above the Earth's
atmosphere, the only sources of image distortion are diffraction
and optical imperfections in the mirrors and telescope structure.

The performance specifications of ST are that the Ritchey-
Chretien optics produce an image of a point source at 6328 Å such
that 70% of the energy will be included within a radius of 0.1 arc
sec. Toward longer wavelengths the performance will rapidly
approach the diffraction limit. Toward shorter wavelengths, the
actual characteristics may not be determined precisely until the
telescope is in orbit. The absolute performance should improve
somewhat at shorter wavelengths as the effects of diffraction
become smaller. However, it is also possible that small scale
telescope imperfections may become significantly more important in
the ultraviolet than they are at red wavelengths.

As of this writing, polishing work on two different candidate
primary mirrors is nearing completion at the facilities respectively of the Perkin-Elmer and Eastman-Kodak companies. In each
case pioneering techniques are being employed. The most recent
indications are that the ultraviolet performance will be more
toward the optimistic rather than the pessimistic side of the range
of possibilities.

One characteristic of the space environment that is very
different from the Earth's surface is that the instrumental point
spread function (PSF) is not only very narrow but also very stable
in time. It will be possible to calibrate the PSF accurately and
to employ sophisticated image restoration techniques to produce

spatial resolution that is somewhat better than the raw data. A
conservative estimate is that the improvement will be at least a
factor of ten better than ground-based observations with respect to
linear spatial resolution (>100 in areal information). Such an
improvement will produce Uranus images of a quality comparable to
superior images of Jupiter before the advent of flyby missions.

Resolving power alone does not determine the total information
content of an image, of course. Contrast is also an important
factor. Uranus is well known to exhibit characteristic absorptions
due to methane (CH_4) which are almost 100% in the strong bands
longward of 10000 Å and which decrease to invisibility near 5000
Å. Ground-based imaging is adequate to show that image pairs in
and adjacent to the strong CH_4 bands show very prominent differential contrast whereas broadband or short wavelength images do not
(J. Westphal, private communication, 1981). One of the first
generation cameras aboard ST will be equipped with appropriate
filters designed to obtain the maximum possible contrast in
atmospheric features on Uranus.

The actual performance of the Space Telescope at any time will
depend on the camera being used then. It will be related to such
parameters as the image size, the detector pixel size and the ST
data handling capacity. Examples will be discussed later in this
paper. The ultimate resolution will be set by the rms jitter in
the pointing control system, which is expected to be 0.007 arc sec.

Range of Accessible Wavelengths

The ST optical system will permit high efficiency throughput
down to 1200 Å, below which the MgF_2 coatings will rapidly suppress
further penetration. The region between this instrumental boundary
and the ground-based cutoff at 3000 Å contains atomic, ionic and
molecular electronic transitions that are important for all objects
of astrophysical interest. Uranus is not an exception.

Below 1500 Å, CH_4 is strongly absorbing. The degree to which
this absorption and the Rayleigh scattering from H_2 combine to give
the reflectivity in this wavelength range will be determined by the

vertical distribution of CH_4 in the Uranian stratosphere.   There is
almost certainly some saturation of CH_4 on Uranus (eg. Danielson,
1977).   The far ultraviolet spectrum offers an excellent prospect
for modelling this situation quantitatively.   No other astronomical
satellite has had the capability of detecting Uranus at such short
wavelengths.   It will be a routine observation for ST.

The other three giant planets beside Uranus all have distinct
stratospheric thermal emission features due to ethane (C_2H_6) at
12 μm and acetylene (C_2H_2) at 13 μm.   Uranus does not (Gillett and
Rieke, 1977).   The difference could be due either to Uranus' having
a cooler stratosphere than the other giants or to lower hydrocarbon
abundances, or both.

These molecules also have electronic absorption features.   In
particular, C_2H_2 has a distinctive series of absorption bands
between 1700 and 1800 Å that have been detected on Jupiter and
Saturn (eg. Owen et al., 1980).   However, Uranus is at the limit
of detectability at these wavelengths (Caldwell et al., 1981) and
Neptune is too faint to have been observed there with previous
satellites.   Spectra of Uranus and Neptune below 1800 Å would be
exceptionally useful in clarifying the statospheric compositions of
the outer two giant planets; both would be easy targets for ST.

The upper limit to the useful wavelength range for the ST
Observatory is longward of 1 mm.   There is little incentive to
investigate this region with ST, because many ground-based radio
telescopes will perform better there.   However, it is expected that
ST will generally do better than ground-based instruments throughout the three decades of infrared radiation from 10^{-4} to 10^{-1} cm
wavelength.

The telescope optics will be at the ambient temperature of
near - Earth orbit.   There will be provision for cryogenically
cooled detectors, although no such detectors are included in the
first generation of instruments.   The absence of the background
radiation from the Earth's atmosphere and the long-term stability
of the PSF will permit ST to perform background subtraction for
infrared photometry by moving the entire telescope to a nearby sky

pointing after minutes or hours of integration on the target, rather than chopping between sky and target at a frequency of tens of hertz, as ground-based telescopes typically do.

Uranus has many unusual infrared properties that could profitably be investigated with the ST. For example, its geometric albedo at 5 μm is 2×10^{-3} (Brown et al., 1981), by far the lowest of the giant planets there. This must be due both to a general absence of particulate scatterers in the stratosphere of Uranus, and to very strong molecular absorbers in this spectral region.

From the ground, the signal from Uranus is so weak and the sky is so bright that it will probably be impossible to perform high resolution spectroscopy to identify the absorbers. From ST, a suitable spectrometer would have a much better chance of success.

To cite another example, photometry of Uranus longward of 10 μm is valuable for determining the thermal structure of the atmosphere of Uranus, but it is extremely difficult to achieve from the ground the photometric accuracy necessary to constrain models usefully. ST could also make improvements here.

In summary, the ST will have an unobstructed spectral range over four decades in wavelength, from the far ultraviolet to the submillimeter range, much of which is totally or partially blocked for ground-based observers.

Limiting Sensitivity

Implicit in the discussion of the preceding section is the realization that the terrestrial sky brightness limits the faintness to which ground-based telescopes can observe. For many aspects of planetary astronomy, this limitation is unimportant because planets are typically bright objects. However, the improvement in limiting sensitivity from space is one of the most important advantages that ST has for many other branches of astronomy, notably cosmology.

The observational limit is, of course, a function of the time which is invested in a project. With the practical assumption that ST integration times will be limited to ten hours, the ST capabilities may be summarized as follows: it is expected to be able to reach at least two stellar magnitudes fainter than the best ground-based telescopes have done in the "transparent" windows of the Earth's atmosphere, and to do very much better than that between the windows. It will reach about eight magnitudes fainter than previous orbiting ultraviolet observatories.

Absence of Scintillation

The rapid variations of the optical paths through the Earth's atmosphere limit astronomical observations of variable events with time scales comparable to that of the atmospheric variations. This scintillation also limits the ability to do accurate photometry of faint targets that are close to bright ones, as the Earth's atmosphere randomly confuses the two signals.

These general limitations restrict a type of planetary obser-vation that has come to be exploited successfully in the past decade - the occultation of stars by Solar System objects. For airless bodies, such events provide information on diameters and shapes that is more accurate that even ST imaging can achieve (eg. Millis et al., 1981)! For planetary atmospheres, differential refraction during the occultation leads to a measurement of the atmospheric refractivity with height, which, together with the assumption of hydrostatic equilibrium, leads to models of the atmospheric thermal structure at very low density levels. However, terrestrial scintillation introduces noise features that can grossly distort the derived models, and which can render intrinsic planetary atmospheric features confusing.

In the space environment of ST, the only variability will be that of the event itself. ST can take advantage of the absence of scintillation to include in an observing aperture only a small portion of the planet being occulted. This would be hopeless from the ground. This advantage could be translated into higher signal-to-noise ratios for specific events, or the ability to observe occultations of fainter stars, which are much more numerous.

Because ST will occupy only a single point in space and time, it will generally be able to observe only occultations by objects with large shadows - that is, the giant planets and their ring systems. Asteroids and small satellites will probably remain for ground-based observers with small, mobile telescopes, who can alter their location to maximize the observability of a particular event.

Occultations by Uranus should therefore form a significant portion of this kind of planetary activity from ST.

THE FIRST GENERATION SCIENTIFIC INSTRUMENTS
Overview

Scientific instruments (SI) are located near the Cassegrain focus of the telescope, behind the primary mirror. There are two general types: "axial" instruments are about the size of a coffin, and have their long axes parallel to the telescope optical axis; "radial" instruments resemble a truncated wedge, are of the same order of size and are located between the axial instruments and the back of the primary mirror. There are four "bays" for each kind of instrument. Three of the radial bays contain Fine Guidance Sensors, which sample the outer parts of the focal surface, nine arc minutes from the optical axis, and provide the information for fine guiding and astrometry. The fourth radial bay samples the middle three arc minutes of the focal surface by means of a pick-off mirror. All four axial instruments have a corner touching the optical axis, behind the radial pick-off mirror. Light therefore enters these instruments only in an off-axis direction, requiring internal corrections to restore image quality.

The Observatory selects the current observing instrument, and particular apertures in that instrument, by means of small pointing manoeuvres of the entire spacecraft.

Wide Field/Planetary Camera

This is the first generation radial instrument. Its development team is led by Professor James A. Westphal of the California Institute of Technology. Its primary purpose is imaging, with some low dispersion spectroscopic capability. It has the widest field of view of any SI, and the broadest spectral range.

The camera has two modes of operation, selectable by a rotation of an internal reflecting pyramid. These modes are named "wide field" (f/12.9; 2.7 x 2.7 [arc min]^2) and "planetary" (f/30; 1.2 x 1.2 [arc min]^2). In each mode, the detector consists of 4 silicon charge-coupled devices. There are 8 such CCDs in the camera, each including 800 x 800 pixels. For each mode, the image is reassembled in the data processing stage into an equivalent 1600 x 1600 image, with no loss of information at the internal boundaries.

Its wide spectral range is made possible by an organic phosphor coating, which converts ultraviolet photons to visual ones, where the intrinsic CCD efficiency is high. The net quantum efficiency exceeds 1% from 1150 Å to 11000 Å, peaking at 55% near 5500 Å. It has the only near infrared imaging capability of the first generation SIs.

Its photometric accuracy is ~1%. Its dynamic range is 10^5, permitting the simultaneous recording of widely different brightness levels in the same field. In normal imaging, spectral definition comes from 48 filters, including methane (CH_4) band filters for imaging the giant planets. There will also be prisms and transmission grating for low dispersion objective spectra, and polarizers at several wavelengths.

Faint Object Camera

This axial instrument, designed and built by the European Space Agency, also has two modes: f/96 (11 x 11 [arc sec]^2) and f/48 (22 x 22 [arc sec]^2). Its field of view is actually limited by the observatory data handling capacity. It will achieve the highest possible spatial resolution and will reach the faintest objects for the first generation SIs. It has much greater speed than the WF/PC in the ultraviolet. The peak quantum efficiency is ~26% at 3000 Å.

It will contain approximately 48 selectable filters, including neutral density and special purpose ones, objective prisms and

polarizers. It will also have a coronographic capability, featuring
a 0.6 arc sec occulting disk that will permit recording objects 1
arc sec apart which differ in brightness by ~ 10 stellar magnitudes.

There will also be a fixed grating spectroscopic mode, permitting observations in three orders: 3600 - 5400 Å; 1800 - 2700
Å and 1200 - 1800 Å at a resolution of $\lambda/\Delta\lambda \sim 2 \times 10^3$. This performance is less than those of the spectrographs discussed below,
but this instrument has the unique feature among first generation
SIs of providing spatial information perpendicular to the dispersion direction. Its entrance slit is 10 x 0.1 (arc sec)^2.

Faint Object Spectrograph

This axial instrument is being developed by a team led by Dr.
Richard Harms of the University of California at San Diego. It
will obtain low and moderate resolution spectra of objects fainter
than those accessible from the ground. It has two operational
modes, with resolutions respectively of $\lambda/\Delta\lambda \sim 10^3$ and 10^2.

It will have a usable spectral range from 1150 Å to 9000 Å,
although its quantum efficiency drops to 1% near 7000 Å and
decreases rapidly toward longer wavelengths. It will also exploit
the ST spatial resolution capability on extended sources, with 10
entrance apertures ranging in size from 0.1 to 4.3 arc sec.

Other features of the FOS include polarization capability
below 3000 Å, and a time resolution for sufficiently bright objects
of 50 µ sec. The observatory data handling capacity will permit up
to 100 exposures/sec for the 512 diode spectra.

High Resolution Spectrograph

This instrument is being developed by a team directed by Dr.
John C. Brandt of the Goddard Space Flight Center. It will provide
the highest spectral resolution of any of the first generation SIs.
There will be three modes, with $\lambda/\Delta\lambda \sim 10^5$, 2×10^4, and 2×10^3.

Its high resolution capability is achieved at some cost in
other parameters. For example, it will operate only below 3200 Å,
and it will have only two selectable spatial resolutions, 0.25 and

2.0 arc sec. However, it will have high time resolution capability
(~ 25 m sec), and it will also have a finite response below 1150
Å, both useful for very bright sources.

Its highest resolution observations will be totally without
precedent in space astronomy. Therefore, astronomical objects
themselves will not be totally adequate as calibrators, and the HRS
will rely on its own internal light sources for spectral and photometric calibration.

High-Speed Photometer

Dr. Robert C. Bless of the University of Wisconsin is the
leader of the development team for this, the mechanically most
simple SI of the first generation. It has two basic purposes:
high time resolution (up to 16 μ sec) and high photometric precision (0.2%). In both these respects, it will be the best performer
among first generation SIs. It can be related to a ground-based
clock with an accuracy of 10 m sec.

Its detectors include four image dissectors which together
cover the range from 1150 Å to 6500 Å. Apertures of 0.4, 1.0 and
10 arc sec will be available.

A fifth detector, a side-looking gallium arsenide photomultiplier tube with peak sensitivity from 6000 to 9000 Å, was added
specifically to enhance the observability of stellar occultations
by planets. The red-sensitive tube is desirable because most stars
in the sky are red, and the light from giant planets can be greatly
suppressed in the red by choosing a filter in a strong methane
band. Since it is the starlight which is the useful signal in such
an observation, the best signal-to-noise therefore typically occurs
in the near-infrared CH_4 bands.

The HSP will also be able to measure linear polarization below
4000 Å. Its dynamic range will be ~ 10^8.

Fine Guidance System

There are three fine guidance sensors, which sample the focal
plane outside the SI fields of view. The light detector in each

sensor is an image dissector/interferometer combination. The
system is used for guiding the observatory, with 2 FGSs required at
any time for this purpose. The third is free to perform astrometric observations.

Each sensor will measure relative positions within its own
field of view (69 [arc min]^2) to within 0.002 arc sec. The magnitude range for astrometric targets (including those observed
through netural density filters) is 4 m_V to 20 m_V. It will be able
to track moving targets.

The ST astrometry science team leader is Dr. W. H. Jefferys of
the University of Texas at Austin.

Future Instruments

The descriptions above refer to instruments selected in a
competition in 1978. It has always been the intention of NASA to
provide for the replacement of these instruments as they become
obsolete or defective. There will probably be an announcement of
opportunity for at least one new instrument before this paper is in
press. Among the obvious areas where the first generation instruments are deficient are the lack of high spectral resolution above
3000 Å, the lack of imaging above 1 μm and the general lack of
infrared capability. These areas will be prime ones on which
second generation instruments will focus.

It should be emphasized that NASA's record in this entire
project is consistent with choosing the best possible science for
ST. The competition for second generation instruments will be
completely open to all who wish to participate, and all will be
given an unbiased evaluation.

SCIENCE MANAGEMENT

In an unprecedented approach to the task of operating the ST,
NASA is in the process of contracting management responsibilities
to the Association of Universities for Research in Astronomy, Inc.
(AURA). AURA has chosen the Johns Hopkins University in Baltimore,
Maryland, to be the location of its Space Telescope Science Insti-

tute, an independent establishment that will run the Observatory.

It will solicit observing proposals from the astronomical community, evaluate those proposals in a peer review system, provide such support as guide star selection for fine pointing, collect and archive the data, and provide facilities for data analysis. It will be the astronomer's prime (and probably only) point of contact with the system.

A related development is that there will be a European coordinating center for ST activities, primarily data analysis support. This will be managed by the European Space Agency. Its site is currently the subject of a competition among several prominent European astronomical centers.

THE LAUNCH DATE

As mentioned in the introduction, the current plan is to launch ST in the first quarter of 1985. This should permit observations to begin in time to participate in two unique astronomical events: the 1985 - 1986 apparition of Halley's comet and the January 24, 1986 encounter of Uranus by Voyager 2 (Stone, this volume).

The current launch schedule is the result of a slippage due mainly to financial problems. Although the schedule has been determined by dollars and not by science considerations, it has, by chance, now been set at virtually the last possible moment that will permit full exploitation of the ST's capabilities for these events. This statement includes recognition of the finite time required to commission the ST after launch, the finite time necessary to assimilate new science, the finite lead time for inclusion of new data into Voyager planning, and pre-perihelion cometary phenomena.

If there is any further slippage, unique science in two areas will probably be lost irretrievably.

Since the primary interest of this volume is Uranus, not Halley, a few considerations of the interaction between ST and Voyager 2 will be given here. It is first noted that at the time

of encounter, Voyager 2 will be travelling nearly parallel to the
rotational axis of Uranus. Therefore, its closest approaches to
the planet, the satellites and the rings will be simultaneous, not
sequential as in previous encounters. Therefore, improved scientific understanding of all the phenomena of the Uranus system
before the encounter will be invaluable in planning for the
precious few hours of greatest opportunity.

It is expected that Voyager's imaging capability will exceed
typical ground-based performance for about thirteen months before
encounter (Stone, this volume). However, ST will exceed ground-
based achievements by an order of magnitude in linear resolution,
and may also have advantage over Voyager 2 with respect to the
location of filters and the intrinsic contrast of atmospheric
features on Uranus. It is therefore possible that the best available information on Uranus will come from ST until such a late time
that further information would not be in time to impact the
encounter plans.

Concerning the detectability of faint rings and satellites, it
is not clear which mission will have the advantage. Starting about
a year from encounter, Voyager will be the best imaging system.
Several months later, after the ST launch, ST will be the best.
Subsequently, as Voyager gets closer, it will regain primacy for a
brief time. If there are currently undetected rings or satellites
of Uranus, their discovery could come from either spacecraft,
depending on how bright they are. It is conceivable that each
spacecraft could take turns discovering progressively fainter
bodies!

CONCLUSION

In the introduction, it was noted that the scientific career
of William Herschel was not limited merely to his observations of
Uranus. Similarly, it is certainly true that observations of
Uranus will form a very small part of the total science program of
ST. However, Uranus will be an important part of that program, and
perhaps one of the most important planetary targets.

Current knowledge of the Solar System is distressingly inhomo-geneous. The more distant memebers of the system are virtual strangers compared to our near neighbours, which are being examined at the microscopic level. Surely it will be impossible to claim a definitive understanding of the whole planetary system until all of its important members become less mysterious. For this reason, it is the opinion of the author that Uranus, together with Neptune and Pluto, should be ranked first among the major planets in terms of ST observing priority.

If the first two hundred years of the study of Uranus are frustrating because it still seems little more than a point source of light, there is some relief in the real prospect that various endeavours, of which the Space Telescope is a leading example, will provide unparalelled progress in the next five.

ACKNOWLEDGEMENT

Presentation of this paper at Bath, U.K., was made possible by a travel grant from the United States National Science Foundation and by support from NASA Grant NSG 7320.

REFERENCES

BROWN, R. H., CRUIKSHANK, D. P. AND TOKUNAGA, A. T. (1981). The rotation period of Neptune's upper atmosphere. Submitted to *Icarus*.

CALDWELL, J. (1981). Planetary science with the Space Telescope. *COSPAR/Advances in Space Science* 1, in press.

CALDWELL, J., OWEN, T., RIVOLO, A.R., MOORE, V., HUNT, G. E. AND BUTTERWORTH, P. S. (1981). Observations of Uranus, Neptune, and Titan by the International Ultraviolet Explorer. *Astron. J.* 86, 298-305.

DANIELSON, R. E. (1977). The structure of the atmosphere of Uranus. *Icarus* 30, 462-478.

GILLETT, F. C. AND RIEKE, G. H. (1977). 5-20 micron observations of Uranus and Neptune. *Astrophys. J.* 218, L141-L144.

LECKRONE, D. S. (1980). The Space Telescope Scientific Instruments. *P.A.S.P.* 92, 5-21.

LONGAIR, M. S. (1979). The Space Telescope and its opportunities. *Quart. Journ. Roy. Ast. Soc.* 20, 5-28.

LONGAIR, M. S., AND WARNER, J. W., editors (1979). *Scientific Research with the Space Telescope*. NASA CP-2111 (IAU Colloquium 54).

MILLIS, R. L. AND 37 others (1981). The diameter of Juno from its
    occultation of AG + 0°1022. *Astron. J.* $\underline{\underline{86}}$, 306-313.
O'DELL, C. R. (1981). The Space Telescope. *Ann. Rev.*
    Astron. Astrophys., in press.
OWEN, T., CALDWELL, J., RIVOLO, A. R., MOORE, V., LANE, A. L.,
    SAGAN, C., HUNT, G., AND PONNAMPERUMA, C. (1980). Observa-
    tions of the spectrum of Jupiter from 1500 to 2000 Å with
    the IUE. *Astrophys. J.* $\underline{\underline{236}}$, L39-L42.
SAVAGE, B. D., COCHRAN, W. D., AND WESSELIUS, P. R. (1980).
    Ultraviolet albedos of Uranus and Neptune. *Astrophys. J.*
    $\underline{\underline{237}}$, 627-632.
SPITZER, L. Jr. (1979). History of the Space Telescope. *Quart.*
    Journ. Roy. Ast. Soc. $\underline{\underline{20}}$, 29-36.

THE VOYAGER ENCOUNTER WITH URANUS

E. C. Stone*

California Institute of Technology

Pasadena, California 91125

The Voyager 2 spacecraft is targeted for an encounter with Uranus in January, 1986. In addition to a brief description of the 11 scientific investigations and the Uranian encounter geometry, the scientific capabilities of Voyager 2 are discussed for the general areas of the atmosphere, the rings, the satellites, and the magnetosphere.

Introduction

In 1977, NASA launched two Voyager spacecraft designed for the study of Jupiter and Saturn and the interplanetary medium at increasing distances from the sun. The trajectory of Voyager 2 was chosen to include the option for continuing on to Uranus and possibly Neptune. The Voyager 2 closest approach to Saturn occurs on August 26, 1981, followed by a closest approach to Uranus on January 24, 1986. The flyby of Uranus has been designed so that the Voyager 2 spacecraft can continue on to an encounter with Neptune on August 24, 1989. Detailed information on the primary mission, on the spacecraft, and on each of the 11 investigations is available in two special issues of Space Science Reviews (21, 75-376, 1977). The general capabilities of the Voyager investigations at Uranus will be similar to those at Jupiter and Saturn which are reported in detail in Science (204, 945-1007, 1979; 206, 925-995, 1979; and 212, 159-243, 1981).

There are 11 scientific investigations on Voyager 2 (see Table 1) involving about 120 scientific investigators. The nominal characteristics of the instruments are contained in Table 2, while Figure 1 illustrates the location of the instruments on the Voyager spacecraft. The four boresighted, remote-sensing instruments (Imaging Science, Infrared Spectroscopy and Radiometry, Photopolarimetry, and Ultraviolet Spectroscopy) share the scan platform which

---

* Voyager Project Scientist

has two axes of articulation in order to provide complete angular coverage.

Due to the greater distance to Uranus, the spacecraft data rates will necessarily be lower than at Saturn. It is currently planned, however, to reprogram the spacecraft to perform onboard image compression and to use an efficient hardware encoder in order to return ~440 images/day, comparable to that achieved at Saturn.

Mission Characteristics

Figure 2 provides an ecliptic plane projection of the trajectories of Voyager 1 and Voyager 2. At the time of the Voyager 2 encounter, Uranus' spin axis will be directed essentially toward the sun, resulting in the appearance of the Uranian satellite and ring system as a bull's eye to the approaching spacecraft. The view normal to the trajectory plane (Figure 3) shows this unique encounter geometry. Figure 3 also indicates that the spacecraft disappears behind Uranus as viewed both from the sun and from the Earth, so that both a solar and an Earth occultation experiment can be performed on the rings and on the planet. The geometry of the radio occultation experiments on the rings and on the planet as viewed from Earth is shown in Figure 4. The approach distances for the spacecraft during these events are indicated in Table 3.

Table 1. Voyager Science Investigations

Investigation Area	Principal Investigator/Institution
Imaging Science (ISS)	Smith/Univ. Arizona (Team Leader)
Infrared Spectroscopy and Radiometry (IRIS)	Hanel/GSFC
Photopolarimetry (PPS)	Lane/JPL
Ultraviolet Spectroscopy (UVS)	Broadfoot/Univ. So. California
Radio Science (RSS)	Tyler/Stanford Univ. (Team Leader)
Magnetic Fields (MAG)	Ness/GSFC
Plasma (PLS)	Bridge/MIT
Plasma Wave (PWS)	Scarf/TRW
Planetary Radio Astronomy (PRA)	Warwick/Radiophysics, Inc.
Low Energy Charged Particles (LECP)	Krimigis/JHU/APL
Cosmic Rays (CRS)	Vogt/Caltech

Table 2.   Instrument Characteristics

Investigation	Nominal Characteristics
ISS	Two Se-S vidicon cameras (f=1500 mm and f=200 mm); Narrow-angle camera: 19 μrad/line pair, 2900 - 6400 Å
IRIS	Michelson interferometer (3.3 - 50 μm) and radiometer (0.33 - 2 μm); 51 cm telescope; 0.25° FOV
PPS	Photomultiplier with 15 cm telescope; 2630 - 7500 Å; 3.5°, 1°, 1/4°, 1/10° FOV; 2 linear polarizers
UVS	Grating spectrometer; 500 - 1700 Å with 10 Å resolution; airglow (1°x0.1° FOV) and occultation (1°x0.3° FOV)
RSS	S-Band (2.3 GHz) and X-band (8.4 GHz); Ultra Stable Oscillator (<4x10^{-12} short-term drift)
MAG	Two low-field (<10^{-6} - 0.5 G) and two high-field (5x10^{-4} - 20 G) magnetometers; 13 m boom; 0 - 16.7 Hz
PLS	Earth-pointing sensor (10 eV - 6 keV ions) and lateral sensor (10 eV - 6 keV ions, 4 eV - 6 keV electrons)
PWS	Sixteen channels (10 Hz - 56.2 kHz); waveform analyzer (150 Hz - 10 kHz); share PRA antennas
PRA	Stepping receiver (1.2 kHz and 20.4 kHz - 40.5 MHz); right and left circular polarization; orthogonal 10 m monopole antennas
LECP	Two solid-state detector systems on rotating platform; 10 keV - 10 MeV electrons; 10 keV/nuc - 150 MeV/nuc ions
CRS	Multiple solid-state detector telescopes; 3 - 110 MeV electrons; ~1 - 500 MeV/nuc nuclei; 3-dimensional anisotropies

Atmospheric Studies

The dynamics, structure, and composition of the Uranian atmosphere will be studied by Voyager 2. One of the important factors for dynamics studies is the expected resolution of the imaging system as a function of time. Figure 5 shows an estimate of the time prior to closest approach during which the planet can be imaged with a given resolution. Also shown is the size of the infrared and photopolarimeter fields of view. About 700 hours before encounter, the resolution of the narrow-angle camera will exceed 700 km/line pair. Since this is the expected resolution

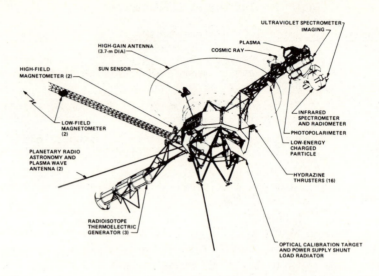

Figure 1. A drawing of the Voyager spacecraft showing the location of the science instruments. The Radio Science Investigation uses the dual-frequency spacecraft transmitters, an Ultra Stable Oscillator, and the 3.66-m parabolic antenna.

from the Space Telescope in orbit about Earth, Voyager 2 will exceed the resolution capability of Space Telescope for a period of about 700 hours during both pre- and post-encounter. The wide-angle camera will exceed the 700 km/line pair resolution during a period ± 4 days around closest approach.

Examples of Saturn images with a resolution of ~700 km/line pair are shown in Figures 6a and 6b, one taken through a violet filter and the other through a green filter. These images demonstrate the scale of cloud structure that may be visible at this resolution, provided of course that the clouds in the Uranian atmosphere have any visible structure.

Near closest approach, the wide-angle camera can be used to advantage as illustrated in Figure 7. The squares overlayed on the views of Uranus indicate the size of the wide-angle field of view which is 7.5 times larger than that of the narrow-angle camera.

An image typical of the wide-angle images to be taken near closest approach is shown in Figure 8. This Voyager image of

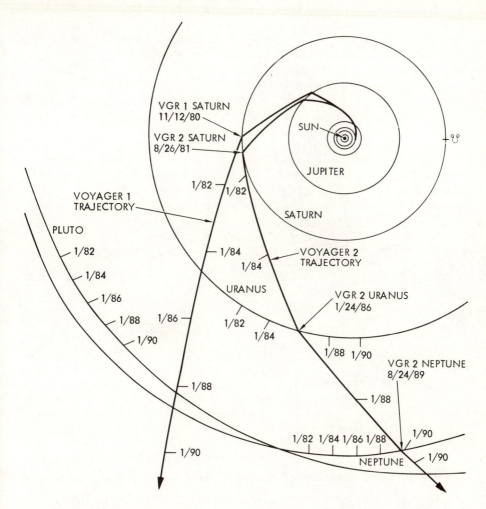

Figure 2. View from normal to the ecliptic plane of the Voyager 1 and 2 trajectories.

Saturn's south polar region was taken from a distance of about 440,000 km when the resolution was about 66 km/line pair. A great deal of structure in the clouds can be resolved from this distance with the wide-angle camera. Even though the lighting level at Uranus is lower than at Saturn, the wide-angle camera will not be smear-limited since less than a two-second exposure with the clear filter will be required. In addition, the wide-angle camera is equipped with a narrow-band filter centered on the 541 nm methane

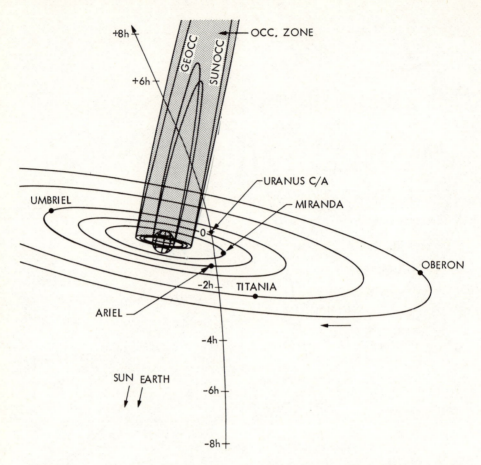

Figure 3. View from normal to the plane of the trajectory for the
Voyager 2 Uranus encounter. The regions in which Earth (GEOCC)
and sun (SUNOCC) occultations occur are shown.

absorption line where some contrast may be expected.

Atmospheric winds can also be derived from a latitude tem-
perature profile of the atmosphere as has been done at Jupiter and
Saturn. The size of the infrared field of view is indicated in
Figure 5.

The study of atmospheric structure and composition will be
addressed by the infrared, radio science, and ultraviolet investigations. The atmospheric temperature/pressure profile will be
derived from the radio science occultation experiment and by

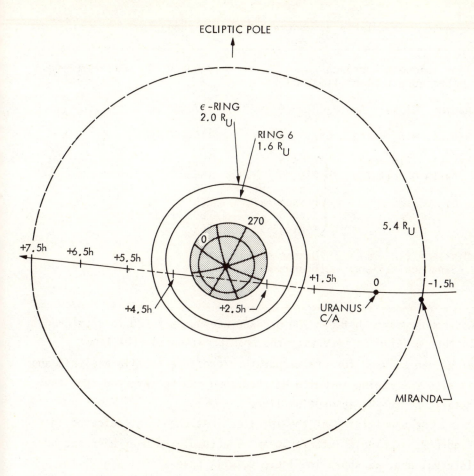

Figure 4. View from Earth of the Voyager 2 Uranus encounter. Only the inner and outermost known rings are indicated. Radio occultation studies will occur as the spacecraft passes behind the rings and the planet.

infrared temperature sounding measurements. The scale heights of H_2 and CH_4 in the upper atmosphere will be determined by the ultraviolet solar occultation measurement, while the same instrument will derive the relative abundances of H and He through airglow measurements.

The infrared measurements are particularly difficult because Uranus is so much colder than Jupiter and Saturn. The IRIS instrument does, however, have adequate sensitivity in the critical

Table 3. Voyager 2 Uranus Encounter

Time, Closest Approach (Spacecraft Event Time, GMT)	January 24, 1986 19:00:00.0
Radius, Closest Approach	107,080 km
Radius, Ring Plane Crossing	115,200 km

Distance (10^3 km) Earth Occultation
$\begin{cases} \epsilon \text{ Ring} & 137 \\ \text{Ring 6} & 155 \\ \text{Uranus entrance} & 186 \\ \text{Uranus exit} & 255 \\ \text{Ring 6} & 280 \\ \epsilon \text{ Ring} & 298 \end{cases}$

Distance (10^3 km) Sun Occultation
$\begin{cases} \text{Uranus entrance} & 177 \\ \text{Uranus exit} & 240 \end{cases}$

---

wavenumber range between 200 and 500 wavenumbers (20 to 50 microns). In this wavelength interval, the pressure-induced S(0) line of H_2 can be used for the temperature/pressure profile analysis and is also the region in which He abundance can be inferred from the details of the H_2 absorption line.

The sensitivity of the infrared instrument is indicated in Figure 9. A single interferogram is accumulated over a 48-second interval and, as shown in Figure 9, will have a 10:1 signal-to-noise at 200 cm^{-1} and will have about 1:1 signal-to-noise at $\sim$350 cm^{-1}. Also indicated are the increased signal-to-noise ratios which are possible with one-hour averaging and with one-day averaging during the encounter phase. With the 24-hour average spectrum, the range should extend to $\sim$450 cm^{-1}, with a signal-to-noise of better than 100 at 200 cm^{-1}. With this sensitivity, it should be possible to derive a temperature/pressure profile between approximately 80 and 1000 millibars (Hanel, private communication).

Although most of the thermal emission from Uranus lies below the 500 cm^{-1} limit of the infrared instrument, a model-dependent extrapolation can be used to estimate the thermal flux at lower wavenumbers. The radiometer portion of the instrument will provide

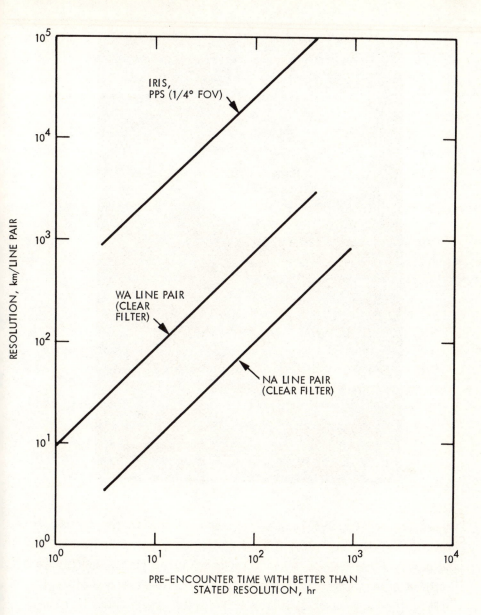

Figure 5. Resolution of the narrow-angle camera (NA), the wide-angle camera (WA), and the size of the IRIS and PPS fields of view as a function of time before encounter.

the phase function of the reflected light, as well as the geometrical and Bond albedo. The ultimate accuracy to which the energy

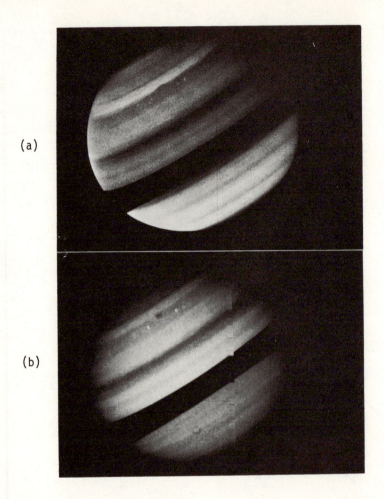

Figures 6a and 6b. Two images of Saturn taken with the narrow-angle camera 25 days before closest approach. The resolution is ~700 km/line pair. Figure a was taken through a violet filter and Figure b through a green filter. Discrete cloud structures are visible at ~40°N. The large dark spot at ~45°N in the green image is elongated in the violet.

balance of Uranus can be determined has not been assessed.

Internal Structure

As we have heard at this conference, models of the internal structure of Uranus depend upon accurate values for a number of observable parameters, such as the oblateness of the planet, the

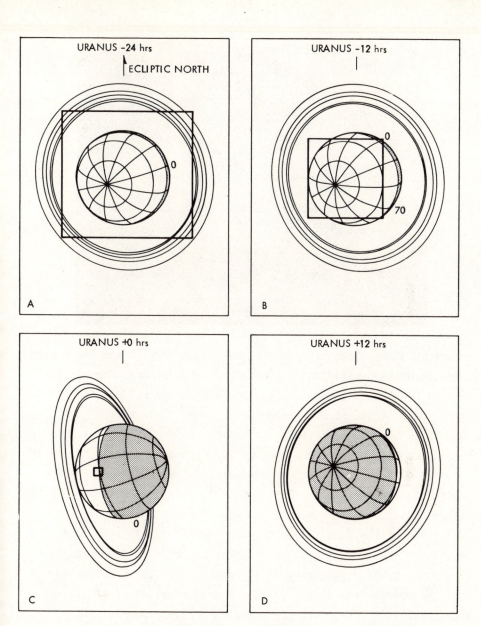

Figure 7.   Views of Uranus from the spacecraft at 12-hour intervals.
The square overlay indicates the wide-angle field of view which is
800 picture elements on a side.

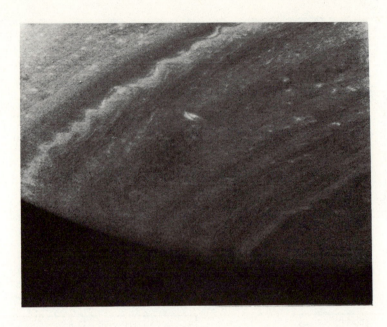

Figure 8. This image of Saturn's south polar region was taken with
the wide-angle camera from a distance of 442,000 km. The resolution is ~66 km/line pair. Waves and eddies are evident in the
large-scale light and dark bands.

zonal harmonics of the Uranian gravity potential, and the period of
rotation. Currently, the best estimate of the J2 harmonic of the
gravitational potential is $3352 \pm 5 \times 10^{-6}$, which has been derived by
Elliot, et al. (Astrophys. J., 86, 444-455, 1981) from the rate of
precession of the eccentric rings around Uranus. Voyager 2 can
provide an independent measure of J2 by an accurate determination
of the spacecraft trajectory during encounter. It is estimated
that the uncertainty of the Voyager 2 determination of J2 will be
approximately $\pm 30 \times 10^{-6}$ (J. D. Anderson, private communication).
Elliot, et al. have also determined a value of $-29 \pm 13 \times 10^{-6}$ for
the J4 gravitational harmonic. Voyager 2 will be able to contribute to the determination of this value through studies of the
eccentric rings. However, the uncertainty of the value of J4
derived from the trajectory information will be $\pm 500 \times 10^{-6}$, too

uncertain to significantly contribute to a more accurate determination of J4.

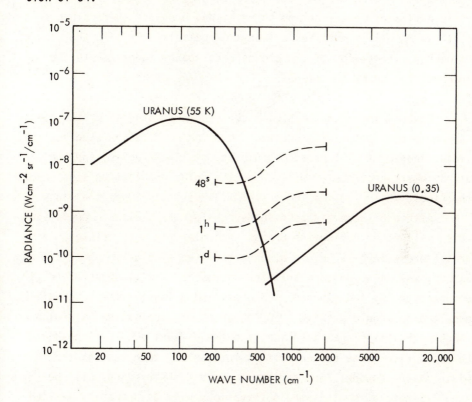

Figure 9. The emitted and reflected radiance levels for Uranus are indicated schematically. The noise-equivalent-radiance for the infrared interferometer is indicated for a single interferogram (48 seconds) and for one-hour and one-day averages.

The radio occultation experiment should provide an accurate measurement of the equatorial radius of the planet and, therefore, provide additional information on that critical parameter.

There are several possibilities for measuring the rotation period for the planet for which current values range from 16 to 24 hours. If Uranus has a magnetic field, it will presumably rotate with the interior of Uranus and it is likely that there will be a corresponding periodicity in the radio emission from the planet

which will be recorded by the Planetary Radio Astronomy receiver.
Time-lapse imaging of the rotational pole of the atmosphere may also
yield estimates of the period of rotation, provided there are discrete azimuthal asymmetries or longitudinal structures in the clouds.
Of course, the period of the clouds need not be identical to the
internal period of the planet.

The Rings

Several investigations will contribute to studies of the rings.
The photopolarimeter will undertake stellar occultation studies, the
radio experiment will perform radio occultation studies, and the
ultraviolet spectrometer will undertake solar occultation studies of
the rings, while the imaging system will directly image the rings.

Although there has been no search for the stellar occultations
which will occur during the Uranus encounter, it seems reasonable
that there will be a number of opportunities similar to those at
Saturn, where the photopolarimeter on Voyager 2 will observe stellar
occultations by the rings of δ Scorpii and β Tauri. The δ Scorpii
occultation should provide ~300 meter resolution of the Saturnian
rings with 10:1 signal-to-noise, and the β Tauri occultation should
be about five times better. It is interesting to note that the
width of the Fresnel Zone for the Voyager 2 observations will be
about 15 meters, so that the observations will be primarily limited
by the 10 millisecond sampling time of the instrument.

The radio occultation studies of the rings will be performed
both at X-band and S-band, that is, at wavelengths of about 3.6 and
about 13 cm. The Fresnel Zones for these two wavelengths will be 3
to 5 km, while the footprint of the X-band antenna beam on the
Uranian ring plane will be approximately 200 km.

It may also be possible to perform a solar occultation study
of the rings using the ultraviolet spectrometer. With the geometry
of our flyby trajectory, the size of the sun corresponds to a distance of about 60 km in the ring plane.

The imaging studies of the rings will occur during both
approach and post encounter. During the approach phase, the resolution may be smear-limited to ~30 km, although a detailed analysis

of the optimum observing strategy has not been undertaken. If the
Jovian ring and the F-ring at Saturn are any indication, the narrow
rings around Uranus may well have a substantial component of very
fine material and may be much brighter in forward-scattered light
than in back-scattered. In that case, the best images of the narrow
rings may occur after closest approach when shorter exposures can be
taken, thereby reducing the smear.

In addition to studying the rings themselves, there will be a
systematic search for shepherding satellites which may be responsible for the stability of the narrow rings at Uranus, much as two
shepherding satellites are responsible for the stability of the F-
ring at Saturn. It is also possible that there will be additional
material around Uranus which can be observed from Voyager 2 but has
not yet been detected from Earth.

The Satellites

Studies of the five known satellites of Uranus will also be a
principal objective of the Voyager encounter. The flyby distances
for each of the satellites are indicated in Table 4. Also indicated
is the range of expected imaging resolution on each of the
satellites. The actual resolution will be determined by the extent
to which spacecraft maneuvers can compensate for the smearing
effects due to the velocity of the spacecraft. Since the detailed
encounter sequence design has not been done, it is unknown how well
such maneuvers can be implemented. However, the technique of image
motion compensation was successfully demonstrated during the Voyager
1 encounter with Rhea. With image motion compensation the resolution on Miranda could be better than 1 km/line pair, comparable to
the best resolution of the Galilean satellites and the intermediate-
sized icy satellites at Saturn.

Other investigations will also contribute to the satellite
studies. The flyby distances are close enough for Miranda and Ariel
to permit infrared measurements of their thermal characteristics.
The photopolarimeter will determine the scattering characteristics
of the icy surfaces as a function of phase angle for both ultraviolet and the far red end of the spectrum. In addition, it may be
possible to search for an atmosphere of one of the satellites if

there is an appropriate stellar occultation.

Table 4. Satellite Encounters

Satellite	Closest Approach (10^3 km)	Resolution* (km/line pair) Geometrical	Smeared
U1 Miranda	16	<1	~15
U2 Ariel	140	3	~30
U3 Umbriel	326	7	~50
U4 Titania	372	7	~50
U5 Oberon	471	9	~50

* Actual resolution will depend upon the extent to which a manoeuver
can compensate for the velocity of the spacecraft relative to the
satellite.

The Magnetosphere

As discussed by Axford at this conference, it is unknown
whether or not Uranus has a magnetic field. If Uranus does have a
magnetic field giving rise to non-thermal radio emissions, the
Planetary Radio Astronomy experiment may provide the first answer
to this basic question. In addition, any one of the fields and
particles instruments can directly detect the presence of a magnetic
field once the spacecraft enters the Uranian magnetosphere. It
should be possible to determine not only whether or not there is a
magnetic field, but also the orientation of the dipole axis with
respect to the spin axis, the strength of the magnetic field, and
the period of rotation.

The Voyager 2 instruments should also provide broad coverage
of the plasma, energetic particle, and wave properties of a Uranian
magnetosphere. Studies of this potentially unique magnetosphere
may be of special significance because it is possible that the solar
wind will be impinging on one of the magnetic polar regions.
Although there have been no detailed calculations of what such an
interaction might be like, it is certainly plausible that the solar
wind plasma could penetrate deeply into the polar region of the

Uranian magnetosphere.

As Axford has pointed out, the Uranian magnetosphere may also be a quiet magnetosphere because of the smaller energy input from the solar wind, in which case the absorption lanes in the trapped radiation environment resulting from the interaction of the satellites and the rings could be quite pronounced. Assuming that the dipole axis is colinear with the spin axis of the planet, the Voyager 2 spacecraft will penetrate into an L-shell of about 4.5 Uranian radii, crossing the magnetic shells associated with the five known Uranian satellites and permitting studies of the satellite-magnetosphere interactions.

Conclusion

This is necessarily a rather preliminary overview of the Voyager 2 capabilities at Uranus. Detailed planning of the sequence of observations will begin in 1984, at which time it will be possible to assess more precisely the potential scientific return from the Voyager 2 encounter with Uranus. It is also likely that new information from continued Earth-based observations of Uranus and possibly from Space Telescope will influence the design of the Voyager 2 encounter sequence. I think it is quite appropriate, however, that in this 200th anniversary of the discovery of Uranus we should be considering the possibility of the first visit to that planet by a spacecraft from Earth.

Acknowledgements

The Voyager Project is being carried out for the National Aeronautics and Space Administration by the Jet Propulsion Laboratory of the California Institute of Technology under Contract NAS7-100.

CONCLUSIONS

URANUS AMONG THE OUTER PLANETS

Tobias Owen, ESS/SUNY, Stony Brook, NY 11794, USA

This conference has offered us a rich and fascinating program. It is most unusual for an astronomical colloquium to include papers on social history, performances of music and readings of poetry, yet all of these apparent diversions have been entirely relevant to our principal topic. In commemorating the 200th anniversary of William Herschel's discovery of Uranus, we are simultaneously celebrating the special qualities of this musician turned astronomer, the extraordinary impact of his discovery on contemporary perceptions of Earth's place in the universe, the very recent growth in our knowledge about the planet itself, and our awareness that we shall soon be learning even more about Uranus from a spacecraft that is presently almost halfway there.

In this short review, I shall not try to summarize the many excellent papers presented at the conference. They are collected in this volume for all to read. Instead I would simply like to emphasize a few themes that have been embodied in these presentations, to try to place Uranus in perspective among the outer planets. The gaps in our present knowledge include some significant unsolved problems on which we can expect real progress in just the next five years.

There seem to be several lessons to be learned from Herschel's discovery of Uranus. The one most often mentioned is the importance of systematic and meticulous observations. We have seen a modern proof of this assertion in the discovery of the planet's rings by James Elliot and his colleagues. These astronomers were not looking for rings, nor was Herschel looking for a planet. But in each case the extent and quality of the observations were sufficient to make the discovery possible.

Another message Herschel left us was the importance of good instrumentation. The telescopes that he produced, as we have heard, were among the best in the world in their optical quality. He insisted on this, and spent great effort achieving it. We find ourselves carrying out this same tradition in the form of large consortia of scientists, rather than as single individuals. In a few years, the most powerful telescope ever built will be sending the data it gathers from Earth-orbit to astronomers on the ground. One of the early targets of this instrument - the Space Telescope - will surely be the planet Herschel discovered. As Dr. Caldwell and Dr. Stone have told us, this telescope will provide the best images of Uranus that we will have for a few months before the Voyager spacecraft reaches the planet in January of 1986.

The last lesson I would like to mention comes from the attitude of other people associated with the discovery. At the time Herschel found Uranus, he was essentially an unknown amateur. He thought he had found a comet, not a new planet. How easy it would have been for the established, professional astronomers to have taken credit for the discovery, since it was up to them to prove that the new object was indeed a planet. So we can admire the generosity of spirit of Nevil Maskelyne, who made certain that full credit was in fact given to William Herschel.

What have we learned about Uranus since Herschel's time? I obviously cannot offer a personal perspective that spans 200 years, but I can at least cover the last 20. Most of what we know about Uranus has been learned very recently. The high quality of this conference is an eloquent testimony to that fact: there are probably more scientists studying Uranus today than the total number who ever spent time seriously thinking about the planet before 1961. At that time, we knew that Uranus had a deep hydrogen atmosphere. Ten years earlier, Gerhard Herzberg had successfully identified one of the pressure induced dipole absorptions of hydrogen in the planet's spectrum as recorded by Gerard Kuiper. This was in fact the first detection of hydrogen in a planetary atmosphere. The strong absorptions seen in the visible spectrum since the early visual studies of William Huggins in the 19th

century had been identified with methane by Rupert Wildt and
confirmed and extended by Arthur Adel and V. M. Slipher in the
1930's. The major spectroscopic puzzle in 1961 was the identification of five regularly spaced lines near 7500 Å, which turned
out to be caused by methane. The only (apparently!) well-known
quantity at that time was the planet's rotation period which, as
we have heard in Dr. Goody's excellent review, was thought to be
10^h 49^m. Clearly if we had held this meeting in 1961, it would
have been a very short one!

To see what we know now, one can consult the Proceedings in
this volume. As a way of underlining some of the important
characteristics of Uranus, it is helpful to draw up a list of
those properties that make the planet unique. For many years, it
has been customary to divide the outer planets into two groups by
bulk properties: Jupiter and Saturn were assumed to be very
similar to each other and to the composition of the primordial
solar nebula, while Uranus and Neptune seemed to be another
similar pair which differed from their giant neighbors by having
much less hydrogen and helium. But we have learned in just the
last few years that these two outer planets are very different from
each other; the papers presented at this conference have strongly
underlined this distinction. A personal list of properties unique
to Uranus is given in the accompanying Table.

Let us briefly go through these in turn. The near-alignment
of the planet's rotational axis with the orbital plane must
produce unusual effects on the circulation and/or chemistry of the
portion of the atmosphere that responds to solar insolation. We
have just seen evidence of a delayed seasonal response in an outer
planet atmosphere in the pictures of Titan returned by Voyager 1.
We may be witnessing some aspects of the reaction of Uranus to its
unique orientation in the microwave observations that have been
collected during the past decade. As Dr. Axford has emphasized
in his comprehensive discussion, the effects of this orientation
on the interaction of Uranus with the solar wind are more difficult
to evaluate, since we presently have no measurement of the planet's
magnetic field. The Voyager 2 spacecraft will provide us with the

necessary data to understand both of these effects when it
encounters Uranus in 1986.

Neptune is now the only outer planet not known to have a
ring system. But the rings of Uranus are distinctly different
from those of either Jupiter or Saturn, being both narrower and
darker. The theory of Peter Goldreich and Scott Tremaine,
advanced to explain these rings, has found observational
verification to first order with the Voyager 1 observations
of Saturn's F-ring. Yet there are still problems to be resolved,
as Drs. Brahic, Dermott, and Elliot have emphasized. We may
anticipate new observational data from forthcoming stellar
occultations, even before Voyager 2 arrives.

Uranus is presently the only outer planet with no known
satellites in retrograde orbits. Since such satellites are often
thought of as captured objects in the systems where they occur,
and since we now have evidence of at least one unattached object,
Chiron, roaming the space between Uranus and Saturn, the absence
of any retrograde body about Uranus is intriguing. Perhaps this
absence is associated with the planet's unique axial alignment.
Or perhaps we should emphasize the word presently in this
discussion, since the search for faint satellites in this system
has not been pushed as hard as it could be. Six satellites were
added to the Saturn system in just the last two years!

But we can be quite certain that there is no large satellite -
nothing as big as the Galilean moons of Jupiter, with Titan,
or with Triton. To try to understand the reason for this anomaly,
we must first find out more about the moons that do exist, to see
how they compare in composition and surface histories with the
other satellites we know.

Uranus is the only outer planet that has not been found to
radiate more energy than it receives from the sun. One could
reconcile this with Jupiter and Saturn in terms of a basic
difference in composition and internal structure. It becomes a
more subtle problem with the awareness that Neptune does have an
internal source of energy. One must look to differences in
formation histories for a possible answer.

The observed variability of the microwave radiation with time may be related both to the orientation of the planet's axis and to the absence of a deep seated energy source. Somehow, one must explain an apparent change in microwave opacity as the pole of rotation is turned toward the observer. We have still only seen less than 25% of the cycle, so even this correlation is not secure. Observations during this decade will provide the necessary leverage.

The upper troposphere of Uranus seems reasonably cloud free. This condition was thought to obtain for both Uranus and Neptune until a few years ago, when distinct "weather patterns" were deduced for Neptune on the basis of observations of periodic variations in the strengths of infrared absorption bands. Nothing comparable has been seen on Uranus. Evidently the temperature lapse rates in the atmospheres of these two planets are distinctly different, a point that is emphasized in another way by the fact that Uranus does not exhibit strong emission bands from ethane and methane in the 8-15μm region. All the other outer planets (including Titan) show evidence of an upper atmosphere thermal inversion this way. Why Uranus does not remains a puzzle. Is it the axial orientation? The absence of internal heat? Some basic difference in atmospheric composition? We simply don't know.

Having emphasized these unique properties of Uranus, I should reiterate that kinship with Neptune is still manifested in that both planets are deficient in hydrogen relative to Jupiter and Saturn. This difference is revealed most clearly in the bulk mean densities of the four objects. There is still controversy - as exhibited in papers presented here - about the methane mixing ratios in the atmospheres, which might be expected to reveal the same effect. This is another area where I think progress can be made before the Voyager 2 encounter.

Dr. Hubbard has raised the interesting possibility that both planets may have formed from comet-like planetesimals and could therefore have atmospheres containing no helium. (Since Herschel at first thought he had found a comet when he sighted Uranus, this evolutionary sequence would probably please him.) An

associated problem is posed by the excess microwave radiation
exhibited by both planets in the 2-10 cm range, as described by
Dr. Gulkis. If Uranus alone exhibited this feature, one might be
inclined to attribute it to the conversion of NH_3 to N_2, as
observed on Titan. But to find it on Neptune too, where an internal
heat source should keep the troposphere mixed, makes this
possibility less likely. A basic difference in the relative
abundances of nitrogen and sulfur might be associated with formation
from comet nuclei. On the other hand, Dr. Belton's provocative
suggestion that Uranus might have a solid surface at a pressure
of just a few tens of bars could allow a photochemical explanation
to remain viable, if he turns out to be correct.

This brief survey should provide some indication of the
progress we have made in understanding Uranus and its place in the
solar system since Herschel's day. We might emphasize how much
is left for us to do by mentioning three areas where there has
been little advance. We still do not know what Uranus really looks
like. We have scattered images of various qualities taken at
different wavelengths, including the excellent set obtained by
Robert Danielson from a telescope carried by a high-altitude
balloon. Good as these Stratoscope images are, they are extremely
limited in both the wavelength and time domains. This is one of
the reasons that we are still uneasy about such questions as the
period of rotation and the presence of clouds and hazes. Given his
emphasis on high-quality telescopes, Herschel might be surprised
that our own planet's atmosphere has proved to be the limiting
factor in obtaining better resolution. He would surely be as
delighted as we are at the prospect of the Space Telescope.

In an entirely different realm, it is sobering to realize
that only one paper at this conference was delivered by a woman.
I think that would have been a surprise to Caroline Herschel.
Devoted as she was to her brother, her own accomplishments would
surely have led her to expect that women would be playing a more
prominent role in astronomy by this time. Her career is a
challenge to us to make such opportunities more available.

Finally, we are still coping with the wonderful cosmogonic problem this planet poses: What combination of processes set it spinning in such an unusual way? It may be that we shall never be able to find a unique answer to this question. But we can expect a rich harvest of data during the next few years from the wide variety of ground-based and near-Earth techniques represented at this conference even before the Space Telescope and Voyager observations are made in 1985 and 1986. So we may harbor the hope that when we assemble again in five years time to discuss this fascinating planet, some real progress will have been made in all three of these areas.

Yet as we anticipate our own interest in the new understanding of Uranus we will have then, we might pause to consider again the impact of Herschel's discovery on the people of his time. We have become rather spoiled with new results because of the many splendid accomplishments of the program of planetary exploration during this last decade. Descriptions of new phenomena or intriguing pictures of distant planets seem to come along every year or two. But in Herschel's day, life did not move at such a pace. Finding a planet beyond those known to everyone was an extraordinary discovery. As just one example, it supplied John Keats with an image expressing his feelings of wonder and delight in his famous sonnet on Chapman's Homer:

> "Then felt I like some watcher of the skies
>
> When a new planet swims into his ken ..... "

I am sure we will have many more opportunities to experience these same feelings as we continue our investigations of this distant and surprising world.

UNIQUE PROPERTIES OF URANUS

(As of March 1981)

Orientation of Rotational Axis

    Meteorology

    Magnetosphere

Nine Dark, Narrow Rings

No Known Retrograde Satellites

No Large Satellite (R > 1500 km)

No Detectable Internal Heat Source

Time-Variable Microwave Emission

No Thick Upper Troposphere Clouds or Hazes

No Detectable High Altitude Emission

CLOSING REMARKS - Professor P.A. Wayman (General Secretary,
IAU). Dunsink Observatory, County
Dublin, Eire.

It is my pleasant task to offer the thanks of the International Astronomical Union to the Royal Astronomical Society and to its President, Professor Arnold Wolfendale, for the convening of this IAU Colloquium No. 60 on the subject of Uranus and the Outer Solar System. It is particularly noteworthy that the opportunity was taken by the Council of the Society to celebrate the second centenary of the discovery of Uranus and to combine this occasion with a colloquium on a subject of special modern interest, with very recent important developments.

By this combination it has been the rare privilege for Fellows of the Society, attending the Spring meeting of 1981, perhaps with a special interest in the history of astronomy and in the work of Herschel, to be present at an international meeting where the speakers on a topic of wide attraction, namely exploration of the solar system, are drawn from among those who have the highest achievements in this fast-developing field. We expect, as indeed is the case, that about half the contributions originate from the United States. Nevertheless the contributors come from eight countries in all and thirteen countries are represented among those attending. Therefore the Colloquium has admirably met the normal IAU criteria as being representative of international activity, as well as being a Society meeting held in what is probably the most charming and attractive of the smaller English provincial cities.

In providing a successful venue, thanks are due to the Vice-Chancellor of the University of Bath for the excellent facilities of the meeting and of the residences, and to Col. John Green for supervision of these facilities. The welcome by Dr. David Parkin and his team of helpers was greatly appreciated as was the preparation of the premises of 19 New King Street as the Herschel Museum

by the Herschel Society, by the National Maritime Museum and by
Mr & Mrs Hilliard. For the whole local programme Professor Davies
and his Local Organising Committee did an excellent job, the musical
occasions being especially attractive.

For the excellent scientific programme of the Colloquium we are
indebted to Dr. Garry Hunt, Chairman of the Scientific Organising
Committee, and his committee members for an interesting series of
reviews and other papers. The life and work of William Herschel,
in the opening session, was of considerable interest, not least
because his contribution to astronomy epitomises in several ways the
successful performance of individual astronomers in modern times.
Herschel saw clearly the need for a type of telescope that was of
superior workmanship and he made it his business to supply that need.
He came to astronomy from another profession late in life; although
he was not trained as a professional astronomer he was a professional
in every general sense of the word, bringing to his work in astronomy
a capacity for assiduous devotion to the task of observation that
has hardly otherwise been matched. With the whole-hearted devotion
of at least one assistant, his sister Caroline, he followed up his
skill at telescope construction with observing zeal of an extraordinary intensity until late in life and he thus stands as a founder
of a tradition of individual initiative and enterprise in modern
observational astronomy that befits him well to be remembered as the
first President of the Society which has been our host at this
Colloquium.

To the scientific programme I have brought no special knowledge,
but I must express my personal thanks to those who took care in
providing reviews of a subject that has advanced very rapidly with
the success of the Voyager missions. This has been a considerable
contribution to a wider appreciation of the new information available and I therefore am able to mark the conclusion of the
Colloquium with the expression of thanks to all who have participated scientifically to make this a successful and memorable meeting.

INDEX